TECHNOLOGY, INNOVATION and POLICY 5

Series of the Fraunhofer Institute
for Systems and Innovation Research (ISI)

W0036753

Knut Koschatzky (Ed.)

Technology-Based Firms in the Innovation Process

Management, Financing and Regional Networks

With 31 Figures
and 28 Tables

Physica-Verlag

A Springer-Verlag Company

Dr. Knut Koschatzky
Fraunhofer Institute for
Systems and Innovation Research (ISI)
Breslauer Str. 48
D-76139 Karlsruhe, Germany

ISBN 3-7908-1021-5 Physica-Verlag Heidelberg

Cataloging-in-Publication Data applied for
Die Deutsche Bibliothek – CIP-Einheitsaufnahme
Technology based firms in the innovation process : management, financing and regional networks;
with 28 tables / Knut Koschatzky (ed.). - Heidelberg : Physica-Verl., 1997
(Technology, innovation, and policy; 5)
ISBN 3-7908-1021-5

Cover design: Erich Kirchner, Heidelberg
SPIN 10629238   88/2202-5 4 3 2 1 0-Printed on acid-free paper

Preface

The Fraunhofer Institute for Systems and Innovation Research (ISI) is one of 47 institutes of the Fraunhofer Gesellschaft, the leading organization for applied research in Germany. ISI performs interdisciplinary research at the interface of technology, economy and science. By the analysis of promising technologies, development of research priorities and monitoring of technology policy programmes, ISI assists decision-making processes in the public and private sectors. This work is mainly contract research.

The institute promotes technical and organizational innovation processes in industry, service sectors, state facilities and private households. Technology forecasting and assessment, regional analyses and innovation services are offered in the fields of production, biotechnology, energy, environment and communication. Interdisciplinary methods of analysis, assessment and forecasting such as Delphi surveys, in-depth interviews, techno-economic indicators, quantitative models and qualitative procedures are utilized in this.

ISI comprises seven scientific departments. This reader highlights research work performed in recent years by the department "Innovation Services and Regional Development". Its aim is not only to illustrate the research spectrum of the department, but to present current research results on development problems, management, public promotion and financing of new technology-based firms (NTBFs) in Germany. For a reader not so familiar with the present situation of NTBFs and the supply of business investment capital in Germany this book provides an excellent opportunity for obtaining a comprehensive overview on the current political and scientific discussion of NTBF promotion in the country.

The contributions are based on research projects financially supported by the Federal Ministry of Education, Science, Research and Technology (BMBF), the European Commission and the Deutsche Forschungsgemeinschaft (DFG), to name - and thank - only a few. During the past years, several staff members of ISI contributed to these projects and supported the research work. We want to thank the authors for shaping their research results in a condensed form and particularly the helping hands in the background whose names are not present in the list of authors. This reader would not have been possible without Anne Ray's translations and Christine Schädel's assistance with editing and the final layout.

Karlsruhe, February 1997

Frieder Meyer-Krahmer
Director

Knut Koschatzky
Head of Department

Contents

Introduction: Technology-Based Firms in the Innovation Process:
Object of Theory and Research
Knut Koschatzky   1

I. The Management of Technology-Based Firms

Development Problems of Small Technology-Based Firms and Ways of
Overcoming Them
Franz Pleschak   11

Marketing in New Technology-Based Firms
Franz Pleschak, Henning Werner, Udo Wupperfeld   31

Crises of New Technology-Based Firms
Joachim Hemer   53

Consulting for New Technology-Based Firms
Marianne Kulicke   81

II. The Financing of Technology-Based Firms

The Promotion of New Technology-Based Firms in Germany
Marianne Kulicke   107

The Financing of New Technology-Based Firms
Marianne Kulicke   125

The Venture Capital Market in Germany
Udo Wupperfeld   149

III. Regional Networks for Technology-Based Firms

Innovative Regional Development Concepts and Technology-Based Firms
Knut Koschatzky   177

Innovation Networks for Small Enterprises
Knut Koschatzky, Uwe Gundrum   203

VIII

Technology and Incubator Centres as an Instrument of Regional Economic
Promotion
Franz Pleschak   225

Methods for Ascertaining Firms' Needs for Innovation Services
Emmanuel Muller, Uwe Gundrum, Knut Koschatzky   245

The Adaptation of German Experiences to Building Up Innovation
Networks in Central and Eastern Europe
Günter H. Walter, Ulrike Broß   263

List of Authors   287

Figures

Figure 1:   Areas of marketing 33

Figure 2:   Distribution of the sample into success categories and
             problem categories 62

Figure 3:   Comparison of average annual turnovers 63

Figure 4:   Relationship between rate of growth and incidence of crises 64

Figure 5:   Comparison of the average annual results 65

Figure 6:   Incidence of crises and type of capital investor 70

Figure 7:   Amount of investment capital that had flowed into inter-
             viewed NTBFs up to the time of interview 71

Figure 8:   Need for support compared with support received 93

Figure 9:   Quality of management support - assessment by NTBFs 97

Figure 10:   Points of application of the two access models designed to
             combat barriers to investing in NTBFs 117

Figure 11:   Net capital requirements of the 336 NTBFs benefiting from
             the BJTU pilot scheme 127

Figure 12:   Content of a business plan 131

Figure 13:   Frequency of inclusion of individual components in the
             business plans of NTBFs benefiting from the BJTU pilot
             scheme 132

Figure 14:   Requirements for investors resulting from the particular
             situation of NTBFs 134

Figure 15:   Planned means of financing the innovation projects of the
             336 NTBFs benefiting from the BJTU pilot scheme 142

Figure 16:   Monetary flows in investment financing 160

Figure 17:   Determinants of innovation in firms 179

Figure 18:   Regional innovation determinants 183

Figure 19:   Approaches to the promotion of regional innovation
             potentials 188

Figure 20:   The Rhine-Main Region 192

Figure 21:   Regional innovation networks 205

Figure 22:   Promotion of joint innovation projects 215

Figure 23:   Innovation centres in Germany 227

Figure 24:   Basic elements of an inquiry to identify the needs of firms
              for innovation support 248

Figure 25:   Indirect and direct identification of the needs of firms 251

Figure 26:   Collection and analysis of available data 254

Figure 27:   Organization of interviews with experts 255

Figure 28:   Organization of interviews with enterprises.................................. 258

Figure 29:   Organizing a questionnaire addressed to enterprises 260

Figure 30:   Concept for help in the process of transition.................................. 279

Figure 31:   Map of Slovenia and Croatia.. 282

Tables

Table 1:   Phases in the life of technology-based firms.................................. 12

Table 2:   Sociodemographic characteristics of the founders of
            technology-based firms... 15

Table 3:   Selected sociodemographic characteristics of the founders
            of technology-based firms in the new federal states 16

Table 4:   Previous employers of founders of technology-based firms
            in the new federal states.. 17

Table 5:   Group of founders of supported technology-based firms in
            the new federal states.. 18

Table 6:   Frequency of participation in partnerships in enterprises
            receiving Phase II support 19

Table 7:   Frequency of elements of customer use targeted by R&D
            projects... 23

Table 8:   Meeting of scheduled targets in the R&D projects of new
            technology-based firms.. 24

Table 9:   Most frequent ways of overcoming financing bottlenecks
            during R&D 26

Table 10:   Time of commencement of market preparations............................ 35

Table 11:   Intended performance spectrum of firms receiving support
            under the TOU-NBL pilot scheme 38

Table 12:   Frequency of competitive strategies followed by new
            technology-based firms.. 38

Table 13:   Sales concept of new technology-based firms 47

Table 14:   Risk factors in NTBFs 58

Table 15:   Characteristics of the 42 firms interviewed.................................. 60

Table 16:   Roles of investment companies in the servicing of NTBFs............ 72

Table 17:   Federal programmes for the promotion of foundations 111

Table 18:   Activities in the process of starting-up NTBFs, and the
            financial inputs and outputs associated with them......................... 128

Table 19:   Long-term development of the German business investment
            capital market.. 151

Table 20:   The German venture capital market in 1995, by financing
            phases.. 153

Table 21:   Sources of capital, by sectors, 1995 154

Table 22:   Types of German business investment companies 155

Table 23:   Main aspects covered by regional surveys as a basis for
              decision about setting up technology and incubator centres 229

Table 24:   Criteria for the regional effectiveness of technology and
              incubator centres 231

Table 25:   Shares of selected fields of technology and areas of activity
              of firms in technology and incubator centres 233

Table 26:   Selected characteristics of the technology base of firms in
              technology and incubator centres 233

Table 27:   Advantages and drawbacks perceived by firms in
              technology and incubator centres 234

Table 28:   Evaluation of the offer of technology and incubator centres
              by firms accommodated in TICs 235

Introduction

Technology-Based Firms in the Innovation Process: Object of Theory and Research

Knut Koschatzky

This reader documents the research profile of the Department of **"Innovation Services and Regional Development"** (IR) of Fraunhofer Institute for Systems and Innovation Research (ISI), Karlsruhe, from the angle of new technology-based firms. IR is one of seven departments of ISI and its work concentrates on new, small and medium-sized technology-based firms, their specific problems in the innovation process and their interactions with their business environment. This includes financing and consulting aspects as well as the providing of information on markets and technologies. Cooperation between individual suppliers of finance and innovation services are of interest, as are regional aspects in the innovation process and in technology development. Finally, the design and effectiveness of public policy compensating for deficits in the supply and demand of innovation resources by means of financial assistance, supportive services and regulative measures intended to help optimize innovation processes and promote an innovative "Mittelstand" are examined.

New, small and medium-sized technology-based firms have specific chances in the innovation process. They contribute to dynamic competition, ensure a wide variety of products, close gaps in the market and offer a broad spectrum of innovative services. On the other hand they also encounter specific barriers associated, for instance, with access to capital and information, with relatively high research and development costs and deficits in management and market experience. The IR Department contributes to forming and securing a competitive, innovative "Mittelstand" in Germany and other European countries by its studies and project monitoring and by evaluations on firm foundations, the German and European investment capital markets, business concepts and innovation management in small and medium-sized firms. It is essential, when analysing industrial innovation projects or conceiving public promotion measures, to regard the innovation process in its entirety, from the development of an idea right through to market introduction. As well as the firm's internal activities, the totality of resource transfers into an innovating firm from its specific environment has to be taken into consideration. For this reason, background factors in the environment of enterprises constitute a further important focus of the work of the Department.

New and small technology-based firms have to rely to a great extent on external partners and institutions when obtaining technological and business information, acquiring technical know-how, in business planning and marketing. The more new technology-based firms - and small and medium-sized firms in general - can succeed in embedding themselves into a regional network of information, consulting, transfer, finance and business relations, the better are their development chances. **Innovation services** are characterized by strong dynamics in their affiliation, forms of organization, content and division of tasks between private, publicly supported and public suppliers. The quality of the instruments and services on both the demand and supply side is very heterogeneous and various forms of differentiation, cooperation and integration are found. IR's research on innovation services concentrates on the development, testing and evaluation of new service offers in pilot projects. These include projects on the industrially-oriented processing of industrial and patent information, and the creation of an expert system to provide technological and market expertises on innovative projects in technology-based firms for German savings-banks.

Small and medium-sized enterprises also play a substantial role in **regional innovation and technology policy**. They constitute a significant target group for public promotion policy, since there is an expectation that they will provide important impulses for the stimulation of regional innovation activities. Networks of enterprises, for instance in the form of research and development cooperations between production and service enterprises, or between enterprises and research institutions, act as catalysts in the exploitation of regional innovation potential. Conversely, a regional supply of innovation-relevant infrastructure and services favours innovation and technical adaptation activities in enterprises. These interactions between firms and their environment are the central topic of technology and innovation-oriented regional research in the IR Department. Central and Eastern Europe constitutes another regional focus for research. The Department supports countries in Central and Eastern Europe by active assistance in the design and realization of modern technology policy. IR brings its experiences in the field of technology policy and technology transfer, innovation financing and service research into studies on **technoscientific structural change in Central and Eastern European countries**. Only if industrial innovation activities and industrially-oriented research receive support from technology, economic and science policy measures can the modernisation of these countries succeed. Together with government representatives and experts from the countries concerned, IR elaborates proposals and measures designed to initiate and expand innovative industry.

It is clear from the individual contributions to this reader that the department's studies, although empirically oriented, are at the same time associated with theoretical concepts and that they aim to contribute to theoretical discussion. The **methodological emphasis** is on sample surveys and total surveys, with qualitative and quantitative data usually being gathered both from written questionnaires and interviews. This primary data collection is then complemented by the utilization of statistical sources, literature and other studies as well as business concepts and development processes of new technology-based firms. Thanks to years of experience in investigating new and small technology-based firms, the department's research and project work can draw on a wide-ranging, sophisticated data base on foundation and development procedures of small enterprises, financing and consulting aspects, venture capital supply in Germany and promotion programmes. All these research areas of the IR Department have a feature in common: the use of descriptive and multivariate statistical methods for the formation of hypotheses and for the analysis and interpretation of data. Various statistical programmes are used for this purpose, notably SPSS. In the field of regional research, too, comprehensive data are available on the innovation activities of manufacturing and service enterprises, on the adaptation of technology and on research and development cooperations in various German and European regions and types of regions. For the analysis of regional technology profiles, a regionally-oriented "High-Tech" classification schema has been developed and tested in the Department. This can be used to link manufacturing and service firms and research institutes to individual product groups or fields of high-technology areas. Patent analysis is also used to gather information on regional innovation and technology potentials. A schema has been developed for project work in Central and Eastern Europe which combines situational analysis with the transfer of methodological knowledge, and is capable of adaptation to the circumstances of individual countries.

Due to the various different focuses of the department's work, there are numerous **link-ups with theory** in different contexts. The enterprise stands the centre-point of research. Thus issues at the **interface between business management and micro-economics** play an important role in the department's work. Research at this level is complemented by analyses of the economics of regional and transaction costs and analyses from the viewpoint of innovation theory. All these theoretical approaches share the understanding of innovation as embracing technical, economic and social change, in the sense of revolutionary and evolutionary changes; they all take a systemic view of innovation processes analogous to Kline and Rosenberg's "chain-linked" model.

Broadly speaking, enterprise-related, business management-oriented studies may have one of three theoretical contexts. Some analyses of firms adopt **innovation management** as their viewpoint and their aim. These have the purpose of identifying barriers within enterprises and formulating management tasks so as to steer innovative activity towards long-term competitive advantages. A second theoretical context is the classification of firms' activities according to paradigms of evolutionary economics (following Nelson and Winter, for instance). Particularly with new technology-based firms (NTBFs), the acquisition of competences and a productive knowledge base is a very significant aspect. Although it is often an innovative idea that triggers a firm foundation, routines first have to be developed by learning processes to ensure the survival of the organization - in this case, the enterprise. The NTBF-oriented research of the department is concerned with these learning processes, and with changes in firms' behaviour resulting from the optimized use of internal resources and reactions to signals in their environment. In the course of their early formation, NTBFs have to pass through numerous learning processes which have less to do with technical competences than with management and marketing. Here, **evolutionary and behaviouristic approaches** offer the possibility of explaining the development of firms in terms of behavioural theory and identifying, through analysis, risk and success factors resulting from deficits in information, coordination, control or hierarchical structure. By contrast, a purely neo-classical approach would not be in a position to identify adequately the asymmetries so frequently found particularly in new and small enterprises, and recognize them not as exceptions in a transparent, intrinsically functioning market, but as the rule for a social institution (the enterprise) interacting intensively with its environment. Closely linked with the evolutionary economic concept is the importance of **trans-**

action costs for the initiation and success of the enterprise. The idea of transaction costs, introduced by Coase in 1937, includes the costs of market exploitation and reflects the influences brought to bear on firms by the organizational form of economic activities in a national economy. With regard to entrepreneurial innovation, networks of enterprises and cooperations can reduce the development and transaction costs of the individual firms. Cooperation with external partners in the innovation process is particularly efficient when the transaction costs are high. For new technology-based firms this situation exists, for instance, in the areas of consulting and finance. Thus the formation and development of innovative networks in the projects of new technology-based firms can also be explained in terms of transaction cost theory.

Regional and Central and Eastern European research work has its foundations in two overlapping spheres of theory. On the one hand, regional development and innovation processes are explained and described by complex regional theories of location, mobility, diffusion, growth and development. Here, studies adopt the **spatial science approach** introduced by Schätzl 1974, which expounds the theoretical explanation of the spatial order of the economy, discusses the empirical recording, description and analysis of spatial processes and the steering of their course towards optimized economic and societal objectives. This also requires the judicious use of innovation and technology policy instruments and measures, with the purpose of increasing the innovation potential of a region and improving regional conditions for innovation. Another way in which the IR Department takes account of the innovative context in its regional research is by deriving investigative hypotheses from the field of **network and institutional theory**. As well as explaining the constitution of transfer networks and structures, institutional theory can also be used, for instance, to explain the formation of innovative networks in general (e.g. Håkansson). According to the concept of the "Innovative Milieu" originally formulated by Aydalot and Camagni, the region provides enterprises with a structural basis for their development. Firms have an interest in integrating into the milieu and they enrich it by the formation of territorial networks. Representatives of this school regard the quality of the innovative milieu in a region as a decisive variable for innovation capability and thus for the economic success of an enterprise. On the other hand, Porter and adherents of the theory of flexible specialization consider demand conditions, industrial clusters and particularly the forms of organization of new industries as determining regional development. According to the latter group, it is not the region itself that is the determinant; rather, the region is shaped by the enterprises locating in it. In this field of tension between theories, some of the department's studies seek to gain current knowledge of regional development processes. Network and institutional theories also form a basis for the analysis of enterprises, in that they highlight the relevance of entrepreneurial innovation activities for regional development, albeit from a different angle.

Although in the practically-oriented contributions of this reader, the links with every individual theoretical approach described here are not explicitly followed-up, this introduction is intended to give a clear idea of the theoretical and methodological spectrum within which the research of the "Innovation Services and Regional Development" Department moves. The contributions that follow reflect the structure of the IR's work and illustrate it by selected examples and discussions on the causes of various problems. The first topic group, dealing with the **management of technology-based firms,** commences with a discussion of the development problems of new technology-based firms from a German view-point and ways of solving them. This overview is followed by chapters addressing some important individual aspects of research work on the development paths of technology-based firms. The first of these considers the importance of marketing for the success of these firms. For new and small technology-based firms in particular, marketing and market introduction of their products and services represents a tough test on the firm's road to development, since management frequently does not perceive this part of the innovation process as ranking high in comparison with research and development tasks, and it is often integrated into the business plan too late. Empirical analyses from Germany of this and other hurdles that can cause the crises afflicting technology-based firms are presented in the following chapter. Since the founders of technology-based firms generally possess few management qualifications and little experience in management, there is a great need for consulting and support in business and commercial aspects. In the course of the firm's development, the intensity of this need may fluctuate and its character may vary. The last contribution to the first subject group presents and discusses the topic of demand and supply in consulting, linking it with analyses of firms.

The second section of the reader centres on the **financing of technology-based firms**. In current discussions on frame conditions in Germany as a location for industry ("Standort Deutschland"), complaints are frequently voiced about the inadequate supply of venture capital and the risk aversion of investors. To illustrate the research work of the IR Department in this area so far, the first chapter in this section deals with promotion programmes and measures targeting new technology-based firms. In Germany, since the mid-1980s, there has been a complex spectrum of actions promoting the foundation and build-up of firms. In previous years the IR Department has been involved in the scientific monitoring of three important pilot schemes of the Federal Ministry for Education, Science, Research and Technology - (BMBF). These were the "Promotion of New Technology-Based Firms" scheme (TOU), the scheme of "Business Investment Capital for New Technology-Based Firms" (BJTU) and the scheme "Promotion of New Technology-Based Firms in the New Federal States" (TOU-NBL). The second chapter in this section of the reader discusses instruments and conditions for financing new technology-based firms and describes the relative values of various financing options in the initiation and development stages of this groups of firms. In the third contribution, the characteristics of the structure of the German venture capital market are presented and starting-points for improving the frame conditions for venture capital in Germany are discussed.

Whereas analysis in first two sections of the reader centres on the management and financing of technology-based firms, the third is oriented towards the regional environment and **regional networks for technology-based firms**. Four contributions are concerned with various aspects of regional research on innovation and technology. First, approaches to the promotion of regional innovation potential are presented from the viewpoint of regional and technology policy concepts for technology-based firms. The second chapter examines the importance of innovation-relevant external relations (networks) for small enterprises and gives examples of how networks of this kind can be formed. The next contribution takes a close look at German technology and incubator centres as a regional policy concept. These centres not only provide firm founders and new technology-based firms with the necessary infrastructure for start-ups; they are also intended to create an environment that is supportive of innovation activities in firms. The fourth contribution present methods which can help regional policy-makers and decision-makers to a more accurate assessment of the need of small enterprises for innovation services. Very often, regional innovation and technology policy is still too supply-oriented, in the sense that promotion measures and the science and technology infrastructure do not examine the needs of enterprises closely enough. The final chapter of the third section covers the transfer of the IR Department's experience to assist the process of techno-economic transition in Central and Eastern European Countries. This contribution deals with the function of innovation networks in the transition process and how they can be formed, with examples illustrating the concrete realization of this concept for innovation and technology policy.

These twelve contributions are intended to give insights into the work of the "Innovation Services and Regional Development" Department of the Fraunhofer Institute for Systems and Innovation Research. They also present, from various angles, current, practically-relevant research results on the importance of small and new technology-based firms for innovation and for economic development. With this reader, the staff of the Department hope to generate impulses for discussion. They would welcome your inquiries, ideas and comments and would be pleased to answer them.

Bibliography

Aydalot, P. (Ed.) (1986): Milieux Innovateurs en Europe. Paris

Camagni, R. (Ed.) (1991): Innovation networks: spatial perspectives. London, New York.

Coase, R.H. (1937): The Nature of the Firm. In: Economia (4), 386-405.

Håkansson, H. (1989): Corporate Technological Behaviour. Co-Operation and Networks. London, New York.

Kline, S.J./Rosenberg, N. (1986): An Overview of Innovation. In: Landau, R./Rosenberg, N. (Eds.): The Positive Sum Strategy. Washington, 275-305.

Nelson, R.R./Winter, S.G. (1982): An Evolutionary Theory of Economic Change. Cambridge Mass., London.

Schätzl, L. (1978): Wirtschaftsgeographie 1: Theorie. Paderborn.

I.   The Management of Technology-Based Firms

Development Problems of Small Technology-Based Firms and Ways of Overcoming Them

Franz Pleschak

1. Introducing the Problem

From initial development to maturity, the life of a technology-based firm traverses several partially overlapping phases (cf. Table 1), each with its own characteristic activities, management tasks and problems. The emphasis of firm management shifts in accordance with the ongoing changes in the evolving character of the firm. For a firm to develop successfully, it has to navigate these successive life phases; problems that crop up have to be consciously addressed and solved. Successful management of these processes enhances the firm's prospects of success.

The following contribution presents some typical problems arising in the initiation and development phases of new technology-based firms and suggests ways of overcoming them. Problems associated with marketing and financing are dealt with in separate contributions of this reader (cf. Pleschak et al.: Marketing in New Technology-Based Firms; Kulicke: The Financing of New Technology-Based Firms).

The development problems of new technology-based firms have already attracted the attention of a number of authors (cf. Unterkofler 1989; Baaken 1989; Dietz 1989; Acs/Audretsch 1992; Pett 1994). The statements in this present contribution are based primarily on an empirical survey conducted by the Department of Innovation Services and Regional Development of the Fraunhofer Institute for Systems and Innovation Research (ISI), Karlsruhe, as a part of its scientific monitoring of the German Research Ministry's pilot scheme for the promotion of new technology-based firms in the old and new federal states. The analysis results given here relating to technology-based firms in the old federal states are taken from Kulicke (1993). Statements relating to the new federal states are taken from current ISI studies (Bräunling et al. 1994; Pleschak/Rangnow 1995; Pleschak et al. 1995).

Table 1:   Phases in the life of technology-based firms

Phase	Activities
1. Initiation phase	
1.1 Idea definition	Entrepreneur analyses own situation Participation in information events Contacts with the environment Reaching agreements within the family Developing and testing out innovative ideas Feasibility studies Evaluation of patents and literature Market research Looking at financing options
1.2 Preparation for foundation	Forming a foundation team Selecting type of legal status Gathering information on promotion possibilities Investigating chances of success Defining business plan together with banks and public institutions Deciding on a location for the firm
1.3 Formal foundation of the enterprise	Registering the business Drawing up a company contract Tax registration Acquisition of resources Business start-up
1.4 Establishing business aims and elaborating a business concept	Analysis of customer requirements Analysis of competition situation Definition of market segment Establishing the necessary date of market entry Project plan for research and development Definition of milestones Costs and profits plan Liquidity plan Identifying need for, and sources of, finance Establishing strategic and operational business goals and growth targets

Table 1 (cont.)

Phase	Activities
2. Development phase	
2.1 Research and development	Performing development work Testing, trials, laboratory models Acquiring pilot and reference clients Patent applications Project controlling Analysis of user experiences
2.2 Preparations for market entry	Building up contact network with customers, business partners, suppliers Advertising, stalls at trade fairs, public relations Building up advisory and service functions Dcfining sales channels to target markets Defining detailed price policy
2.3 Preparations for production build-up	Ascertaining need for production resources and production personnel Securing investment financing Qualification of production personnel Establishing division of labour and cooperation Analysis of production costs
3. Market introduction and production build-up	Making the investments Introduction of organization systems First sales Analysis of customer reactions Communication activities Offers
4. Growth phase	Expanding range of product and services offered Acquisition of new target markets and market segments Extending sales network Expanding production capacity Consolidation investments Further definition of organization systems Growth of turnover and workforce
5. Consolidation phase	Stabilizing customer and supplier relations Reviewing organizational procedures Stabilizing permanent workforce Ongoing R&D activity Stabilizing business functions Consolidating business situation Consolidating relations with investors

2.   Problems in the Initiation Phase

The central problem of the initiation phase of a technology-based firm is to work out a business plan for the enterprise. Only if this business plan promises success, both in technical and in economic terms, is there any justification for the act of founding an enterprise. Elaborating a business plan of this kind makes great demands on the knowledge and experience of the founders. Besides being the key figures in providing the technical solution to a problem as defined by customer need and customer demand, they also have to deal, in their capacity as firm founders, with all the concomitant problems of feasibility, productivity and liquidity.

Despite being experts with extensive research and development (R&D) know-how, the founders of technology-based firms do not generally possess the **prerequisites** for dealing with this complex constellation of problems. Why this is so emerges clearly from the following points:

• Although the majority of founders of technology-based firms have many years' experience of R&D in an enterprise, they have rarely accumulated business knowledge or business know-how. This applies to founders in both the new and old federal states, but even more to the new. Thus, approximately four percent of East German founders have sales experience and only seven percent have commercial experience (cf. Table 2).

• Just under 30 percent of the 479 East German founders in the analysis had no experience of working in a firm. Observed over the period from 1990 to 1994, the proportion of founders of this type is on the increase (cf. Table 3). Although these founders have also been engaged in R&D, they have not been working in enterprises but in universities or non-university R&D institutions. Just under half of the founders in East Germany come from these institutions (cf. Table 4). The ratios in the old federal states are typically quite different: 62 percent of founders come from industrial firms and only 21 percent from universities and research institutions.

Table 2:   **Sociodemographic characteristics of the founders of technology-based firms**

Characteristics of founders	Old federal states ("West" Germany) (n=333 enterprises)	New federal states ("East" Germany) (n=212 enterprises)
Incidence of firm founders with experience of industrial firms in % (multiple entries possible)		
• in R&D	49	61
• in manufacturing	25	61
• in sales	20	14
• in other commercial departments	11	4
• no experience in industrial firms	no input	28
Proportion of founders with management experience in %	37	42
Average no. of years of professional experience	11	12
Average no. of years in R&D work	no input	10
Average age of founders at the time of foundation in years	36	40

Source: Kulicke 1993:33; Pleschak/Rangnow 1995:19

- Over 90 percent of founders in the new federal states have a techno-scientific training; as many as 42 percent have demonstrated their capacity for independent scientific research by gaining a doctorate degree (32 % in the old federal states). The high proportion of founders with a doctor's degree in the new federal states is a consequence of the dissolution and re-structuring of academic institutes and R&D institutions, and the cuts in the number of posts in higher education. Employees who were interested in the idea of a foundation took this chance to build up a firm of their own. Less than two percent of the founders are trained in economics.

Particularly in the new federal states, these basic characteristics give rise to **problems relating to the business aspects of the firm concept**. A lack of business experience is a potential source of danger that may result in unrealistic or inadequate business plans. Not only does this hinder founders in their negotiations with investors; it many also have far-reaching consequences for the economic survival of firms. Founders without any business experience run the risk of inadequate "entrepreneurial action". These founders first have to develop from "researchers" to "entrepreneurs". Thus the business qualifications of founders and the advice they receive when putting together firm concepts rank very high as success determinants in the founding of enterprises.

Table 3: **Selected sociodemographic characteristics of the founders of technology-based firms in the new federal states**

Characteristics of founders	Founders				
	1990/1991 (n=141)	1992 (n=156)	1993 (n=108)	1994 (n=74)	Total (n=479)
Frequency of firm founders' experience of industrial firms in % (multiple entries possible)					
• in R&D	73.7	58.3	53.7	48.6	60.5
• in manufacturing	14.2	16.7	9.2	16.2	14.2
• in sales	2.8	5.8	1.9	4.1	3.8
• in other commercial departments	1.4	7.7	7.4	14.9	6.9
• no experience in industrial firms	12.1	25.6	41.7	40.5	27.6
• no statement	8.5	9.6	0.9	1.4	6.1
Proportion of founders with management experience in %	41.8	38.5	42.6	50.0	42.1
Average no. of years of professional experience	11.0	11.1	11.6	13.8	11.6
Average no. of years in R&D work	10.1	9.3	10.9	12.6	10.4
Average age of founders at the time of foundation in years	38.5	39.8	39.3	40.3	39.5

Sources: Pleschak/Rangnow 1995:19

In research on firm foundations, there is a consensus that the personality and characteristics of the founders exercise a decisive influence on the success of their enterprises. As well as the business skills specifically involved in founding a firm, the factors of management skills, knowledge of the sector, non-conflicting interests within the family, good health and qualities such as vitality, ability to make contacts and a good personal reputation should also be present.

Table 4:   **Previous employers of founders of technology-based firms in the new federal states (shares in %)**

Previous employer	Founder			
	1990/92 (n=297)	1993 (n=108)	1994 (n=74)	Total (n=479)
In the new federal states employed in	82.8	82.4	75.7	81,6
• research institutes / academic institutes	17.8	24.1	20.3	19.6
• universities / higher education	24.6	33.3	29.7	27.3
• enterprises	40.4	25.0	25.7	34.7
In the old federal states employed in	7.7	12.0	21.6	10.9
• research institutes / academic institutes	0.3	2.7	4.1	1.5
• universities / higher education	1.7	6.5	5.4	3.3
• enterprises	5.7	2.7	12.2	6.1
Other	2.0	4.6	2.7	2.7
No statement	7.4	0.9	-	4.8

Source: Pleschak/Rangnow 1995:14

The high proportion of founders from universities and non-university R&D institutes raises the question of how these institutes can motivate potential founders and give them the ability to found an enterprise. Further qualification programmes should smooth the way for assistants, scientific employees and doctoral students to acquire an entrepreneur personality. In order to dismantle barriers, counter obstacles and promote the wish to initiate a foundation project - in other words, in order to make people willing to found firms - potential founders have to be supplied with sufficient relevant information and given the ability to found (Pett 1994).

Technology-based firms are often founded by a **team**. In 73 percent of the 212 firm foundations analysed in East Germany, several people with ideas for a foundation (key persons) had got together to found an enterprise. Since in a further eleven percent of foundations a single key person sought other partners, only one in six firm foundations can be considered as true "single" foundations. The proportion of team foundations is considerably higher in the new federal states than in the old, where team foundations make up only 38 percent of the 333 foundations analysed. Traditions of thought and work in East Germany may be regarded as a reason for the higher proportion of team foundations there. In East Germany too, however, the proportion of foundations by single entrepreneurs is tending to increase because the period in which whole teams went self-employed and founded their own enterprises due to re-structuring and dissolution of existing institutions has now gone by (cf. Table 5).

Table 5: **Group of founders of supported technology-based firms in the new federal states**

Characteristics of group of founders	Foundations				
	1990/1991 (n=49)	1992 (n=67)	1993 (n=56)	1994 (n=40)	Total (n=212)
Share of "team" foundations in total foundations (as evidenced in applications for support) in %	87.8	68.7	73.2	60.0	72.6
Average number of applicants	2.9	2.3	1.9	1.9	2.3
Share of "single" foundations in total foundations (as evidenced in applications for support) in % of which	12.2	31.3	26.8	40.0	27.4
• share of "single" foundations with participating partners (as evidenced in support applications)	2.0	14.9	10.7	15.0	10.8
• share of "true" single foundations	10.0	16.4	16.1	25.0	16.5

Source: Pleschak/Rangnow 1995:21

Team foundations have proven advantages. However, they can lead to problems if the personal relations are dominated by scepsis, if the responsibilities and tasks are not clearly delineated, or if an appropriate division of labour and a suitable form of organization are not found. Dissention arises within the team if the individual founders have different perceptions of their entrepreneurial function, if they attach different priorities to individual aims, or if extremes of "superfluous luxury" or "false economy" occur. The advantages of a team foundation are lost if it is not interdisciplinary, consisting only of technical experts, without members who can contribute business and management experience. An interdisciplinary team is a prerequisite for being able to provide customers with solutions to complex problems.

It is a feasible strategy for founders of technology-based firms, if they themselves have insufficient equity capital, to bring in other partners as participants and at the same time as investors in the **partnership**. Although some East German founders held back in this respect at the time of foundation because they want to "make their own decisions and be independent", or because they are afraid of a knowledge drain, half of the 212 enterprises analysed subsequently extended their circle of investors to include other actively involved or external partners. During the period 1990 to 1994 the proportion of East German investors in partnerships showed an upward trend (cf. Table 6). The main motive for including more partners in the firm was to bring in knowledge to support the process of founding and development and the management of technology-based firms. This consideration was the motive in two thirds of the enterprises.

Table 6:   **Frequency of participation in partnerships in enterprises receiving Phase II support (in %)**

Enterprises with	Frequency of participation				
	1990/1991 (n=49)	1992 (n=67)	1993 (n=56)	1994 (n=40)	Total (n=212)
East German partners	10.2	19.4	25.0	30.0	20.8
West German partners	28.6	29.9	25.0	12.5	25.0
East and West German partners	2.0	3.3	3.6	5.0	3.3

Source: Pleschak/Rangnow 1995:26

In the course of the firm's development, founders come round to the idea of **expanding their number of partners**, because

- it becomes necessary to finance further development work,

- strategic R&D, manufacturing and sales alliances have to be formed,

- it is not possible to build up production entirely from their own resources,

- favourable conditions are created for gaining orders,

- an input of knowledge and experience into the firm becomes necessary,

- financing bottlenecks have to be overcome.

Financing presents a particularly serious problem when founding an enterprise. Even if founders make use of promotion programmes such as the TOU pilot scheme in the new federal states, many of them have difficulties in finding a house bank willing to finance their own stake in the firm. This was true for 59 percent of a sample of 46 East German enterprises interviewed. 17 of the 46 (i.e. 37 %) changed their house bank in order to secure the financing.

Problems arise in **cooperation with banks** at this stage in the development of the firm, because

- banks are not sufficiently familiar with the available promotion measures,
- bank advisors change frequently and are not fully aware of the specific factors involved in setting-up technology-based firms as opposed to firm foundations in general,
- banks are shy of the risks associated with new technology-based firms or feel that they cannot make a valid assessment of the firms.

However, problems may also arise because

- founders are not always able to "sell" their concept convincingly enough to the bank and their personality is not sufficiently confidence-inspiring.

Right from the start, i.e. from foundation onwards, the management of new technology-based firms should build up an open relationship of mutual trust with the house bank, so that the bank can follow-up the development of the enterprise and gain an understanding of the innovation process from the inside. In this way, the bank can form an objective idea of the existing technical and market risks.

The financial basis for new technology-based firms improves if, in addition to R&D projects, the **product and service programme** provides for turnover from other products and services right from the start. This not only facilitates financing and improves liquidity, but also brings with it the advantages of customer and market experience, feedback from experience, image build-up and multivalent use of investments in manufacturing technology. Cooperations and relations with suppliers are established and manufacturing know-how is gained. The R&D projects should be coordinated with the other products or services, in order to create synergies for development, production and sales. When distributing potentials among the various departments of the enterprise, care should be taken to ensure rapid, trouble-free work on the R&D projects, while at the same time making sure that the survival of the firm is not endangered if the results of R&D projects do not meet the time schedules for market maturity and production start-up.

3. Problems in Research and Development

The R&D projects of new technology-based firms are oriented towards new products, new processes and new software. An analysis of 212 concepts for enterprises in East Germany shows that 75 percent of the R&D projects are concerned with new products; of these, 29 percent are product developments, 33 percent are complex product and process developments and 13 percent are complex product and software developments. Eleven percent of the R&D projects are process or software developments. New processes as the starting-point for new products make it possible to find new product functions and open up new product applications. This is also true for R&D projects in which product development and software development are integrated.

The R&D projects correspond to current trends in technological development. This is demonstrated by the following features: a higher share of microelectronics, the integration of various different technologies, the systemic character of the solutions, the combination of hardware and software, the offer of services covering the whole product life cycle. The projects include industrial basic research. Their complexity is an indirect expression of a high innovation requirement. Complexity can bring chances if the complex options are modular in design, thus satisfying various different user requirements and enabling all customer demand to be met from one source. In some cases, it may also be possible to put partial options on the market before the complex system has reached market maturity, thus gaining feedback from customers which can have a positive impact on the development process. On the other hand, increasing complexity is generally associated with higher development costs and longer development times, which may overtax the resources of small or new enterprises.

From in-depth interviews with 46 founders it emerged clearly that approximately 40 percent of the innovative ideas had their **origin** in the analysis and prognosis of possibilities of technological development, and about one quarter originated from analysis of market and customer requirements. Just under 40 percent of the founders emphasized that the interaction of technological development possibilities and market or customer requirements in triggering innovation is so close that a definite "pull" or "push" categorization of innovations is not possible. The high proportion of ideas resulting from technological development is understandable in view of the fact that many of the founders come from universities and R&D institutes. In this approach to R&D projects, existing potential is put to good use and the technical risk factor is relatively low, but on the other hand the market risk is correspondingly higher. First, a market has to be found for the new technical solutions and this pushes up the time and expense of market introduction. Approximately 90 percent of the 46 founders interviewed have already worked for several years in the techno-scientific field in which their current R&D project is situated. From a technical viewpoint this is a favourable basis for successful R&D, since the abundance of knowledge from experience excludes various sources of R&D risk. If the R&D project is primarily based on market analysis and customer requirements, however, the risk of encountering difficulties in technical realization is higher, but the prospects of success on the market are better.

Customer and market orientation are the decisive criteria for the quality of the specification "checklists" of the R&D projects. Otherwise there is a danger that technically-oriented founders will frequently "fall in love" with their development projects, but in doing so will miscalculate the demand and will not adjust sufficiently to the economic dictates of market forces. The criterion for evaluating an innovation should always be its usefulness to the customer. If R&D projects are derived one-sidedly from technical development possibilities, there is a danger that the development will not be related to the requirements or wishes of the customer. An important indicator for the success potential of R&D projects is the **customer use** targeted by the technical development.

Almost all enterprises give the wish to offer customers higher quality than their competitors as the aim of their R&D for new products, processes and software developments (cf. Table 7). Improved technical parameters, integration of functions, a broader area of application, greater reliability and flexibility, higher performance, etc., are concrete expressions of quality targets. It is significant that in about 40 per-cent of the R&D projects in the new federal states, the new products and processes are intended to satisfy totally new needs. In these R&D projects, technical solutions may be realized which are based on new technological principles and designed for new application situations. This expresses the high level of innovation in the R&D projects.

Table 7: **Frequency of elements of customer use targeted by R&D projects (in %; multiple entries possible)**

Elements of customer use	Old federal states (n=93 enterprises)	New federal states (n=212 enterprises)
Improving quality	72	84
Reducing costs	40	64
Increasing productivity	n.e.	42
Meeting new needs	10	40
Increasing flexibility	45	35
Ecological usefulness	n.e.	24
Giving a leading edge over competitors	n.e.	23
Social usefulness	n.e.	19

Source: Kulicke 1993:84; Pleschak/Rangnow 1995:29

About one quarter of enterprises emphasized that their customers gain a lead on their competitors by the innovativeness of the solution. This relatively low fre-quency leads to the conclusion that time is underrated as a determinant for effi-ciency. This also emerges clearly from a written questionnaire on **R&D goal at-tainment** addressed to new technology-based firms. The results of this confirm that a proportion of the firms did not fulfil the final goals of their project at the time originally planned (cf. Table 8). Whereas, according to the managers of the young enterprises, there was a high level of goal fulfilment in the technical and economic goals for the new products or processes, this was not true of the goals relating to the development schedule. About 50 percent of the firms did not keep to their time schedule for the R&D projects, and about 20 percent exceeded their planned devel-opment costs. The factor behind this tendency to exceed schedules is clearly the problematic attitude associated with new technology-based firms which strive to attain the technical goals for products "at all costs", even if this means allowing for longer development times and higher development costs.

Table 8:   **Meeting of scheduled targets in the R&D projects of new technology-based firms (n=number of firms)**

Target element	Firms meeting target	
	number	**percent**
Kept to time schedule for R&D projects (n=66)	34	52
Kept within planned development costs (n=66)	52	79
Achieved targeted technical parameters (n=67)	62	94
Adhered to price calculation for new product or process (n=65)	54	82

Source: Pleschak et al. 1995:8

Some of the causes for non-fulfilment of goals lie in the normal risk attaching to R&D projects; others, however, are the consequence of insufficient experience in the planning and management of projects. It is here that starting-points can be found for influencing firms towards more realistic project planning.

36 enterprises made concrete statements about their **reasons for non-achievement of targets**. Reasons frequently mentioned were:

- modification of the R&D projects due to changes in economic or technical conditions (22 % of enterprises),

- changes in cost or price situation (22 %),

- non-foreseeable technical problems (19 %),

- delays in practical testing (8 %),

- lack of capacity due to expansion of product programme (8 %).

The statements of the great majority of managers to the effect that the targeted technical parameters had been achieved need to be put in perspective, since 57 of the 66 managers questioned emphasized that, after completing the R&D project originally planned, further work was necessary in order to be able to market the new products or processes successfully. The most frequent reasons for this were:

- Extension of technical goals, also further development and increased complexity (32 % of firms),

- Acquisition of licences (25 %),

- Carrying out trials and tests (23 %),

- Customer-specific adaptations (22 %),

- Drawing up development results (16 %),

- Opening up new fields of application (11 %).

On the one hand, these reasons express the continuous advance of technical development, necessitating further development and the perfecting, adapting or maturing of products or processes. On the other, in some cases they reflect typical constituents of innovation processes which often receive insufficient attention during project planning. This applies, for instance, to performing long-term investigations, to trials and tests, authorisation and licensing procedures and customer-specific adaptations. In some circumstances, these may incur substantial risks for new technology-based firms.

Unscheduled R&D leads to **financial bottlenecks.** These may arise because of a need for capital to build up the firm which was not yet recognized during planning, or because the assumptions on which planning was based do not come to pass, e.g. regarding turnover from other products or processes, the amount of R&D work needed or the income from the marketing of parts of the R&D results. Even starting market introduction and production build-up earlier than foreseen in the project - i.e. stepping-up the parallel performance of activities in the interests of a faster market impact - leads to a need for capital which was not yet foreseeable as such in the finance plans. The most frequent causes for financing and liquidity bottlenecks are:

- development times are longer that foreseen,

- R&D costs go up,

- turnover from other products or services is lower than planned,

- large orders outside the supported R&D project have to be financed,

- higher costs of market preparation,

- earlier start of production build-up,

- no possibility to market preliminary results or modules of the supported R&D project,

- restrictive attitude of the house bank.

Ways in which new technology-based firms overcome financing bottlenecks are clearly shown in Table 9. The search for R&D promotion measures tops the list.

Table 9: **Most frequent ways of overcoming financing bottlenecks during R&D (multiple entries possible; n=46 enterprises)**

Ways of overcoming financing bottlenecks	Frequency in %
Making use of other promotion measures	33
Increasing turnover from other products and services	26
Exploiting credit in current account	24
Putting off tasks till a later point, or slowing down the development schedule of the firm	24
Bringing down costs (wage reductions, dismissals, etc.)	17
Taking on additional credits from investment partners	15
Increasing the credit in current account	13
Taking on additional bank credits	7

Source: Pleschak 1995:18

An analysis of 212 concepts for enterprises in the new federal states showed that more than 30 percent of the supported firms already had patents, which had mostly been acquired by arrangement with the former employees of the founders. One-fifth of the firms had already applied for new patents in relation to their R&D project and over 40 percent of firms had the intention of doing so in the course of their R&D. The fact that the proportion of firms wishing to patent their R&D results increases from year to year should be viewed in a positive light. It reflects not only the innovation content of the R&D projects, but also a growing awareness of the economic importance of patents. Discounting results that are non-patentable, 15 percent of the firms consider it neither necessary nor useful to aim to patent their techno-scientific results. The reasons for this are: the firms' lack of economic strength, their low market shares, narrowly defined market segments, non-existence of competition.

The new products and processes of technology-based firms in the new federal states are mostly in the following technological fields: metrology (16.8 % of firms), medical technology (11.9 %), process technology (9.9 %), software tools (9.4 %), manufacturing technology (8.9 %), microelectronics (8.4 %). Information transmission and processing technologies predominate, together with physical processes. By comparison, far fewer new products and processes are based on mechanical or optical processes or electronic solutions. Chemical processes and energy technologies are scarcely represented at all. If broken down according to sectors, electrotechnics, precision engineering, optics and electronics predominate.

Not all fields of technology are equally suitable for new technology-based firms. Fields in which large R&D teams are a prerequisite, where the risk is non-calculable or which generate a very high capital requirement for production prerequisites, can scarcely be considered by new technology-based firms. Small firms have advantages in sectors where the capital intensity is low and where their innovation strategies can be clearly differentiated from those of large enterprises. Innovation opportunities are most favourable for small enterprises in the early phases of the product life cycle, and also in situations where the knowledge necessary for the innovation is generated externally to the main producers, at the interfaces between technologies and in interdisciplinary tasks. Situations in which there are no substantial barriers to market entry or where the product has a unique selling proposition are advantageous. Customized developments and manufacturing as well as market niche strategies are typical.

4. Conclusions

From the studies, it is possible to derive success factors for technology-based firms. In the initiation phase, these include: having founders with business qualifications; forming networks with investors and innovation institutions in the regional environment; the systematic conceptualization of a business concept and a financing concept; making use of consulting when planning; and cultivating good working relationships within the team of founders and among investing partners.

Success factors in research and development are: customer and market orientation; using project management techniques for the planning, organization and control of R&D; using checklist planning; keeping the time factor under control; running R&D in parallel with preparatory market and sales activities; building up networks with customers, suppliers and sales intermediaries; patenting R&D results; cultivating a creative working atmosphere. The founders of technology-based firms should conduct their management tasks in such a way as to ensure maximum impact of these success factors.

5.   Bibliography

Acs, J.Z./Audretsch, D.B. (1992): Innovationen durch kleine Unternehmen. Berlin.

Baaken, T. (1989): Bewertung technologieorientierter Unternehmensgründungen. Berlin.

Bräunling, G./Pleschak, F./Sabisch, H.: (1994): Chancen und Risiken vom im Modellversuch TOU-NBL geförderten jungen Technologieunternehmen. 3. Analysebericht. ISI: Karlsruhe, Dresden.

Dietz, W. (1989): Gründung innovativer Unternehmen. Wiesbaden.

Kulicke, M. u.a. (1993): Chancen und Risiken junger Technologieunternehmen - Ergebnisse des Modellversuchs "Förderung technologieorientierter Unternehmensgründungen". Heidelberg.

Pett, A. (1994): Technologie- und Gründerzentren: Empirische Analyse eines Instruments zur Schaffung hochwertiger Arbeitsplätze. Frankfurt am Main

Pleschak, F. (1995): Finanzierungsprobleme von im Modellversuch TOU-NBL geförderten jungen Technologieunternehmen. Study (internal material). ISI: Karlsruhe, Dresden.

Pleschak, F./Rangnow, R. (1995): Ergebnisse des BMBF-Modellversuchs "Technologieorientierte Unternehmensgründungen in den neuen Bundesländern der Jahre 1990 bis 1994. 7. Analysebericht. ISI: Karlsruhe, Dresden.

Pleschak, F./Sabisch, H./Wupperfeld, U. (1994): Innovationsorientierte kleine Unternehmen. Wiesbaden.

Pleschak, F./Werner, H./Wupperfeld, U. (1995): Marktbewährung geförderter junger Technologieunternehmen. 8. Analysebericht. ISI: Karlsruhe, Freiberg.

Unterkofler, G. (1989): Erfolgsfaktoren innovativer Unternehmensgründungen. Ein gestaltungsorientierter Lösungsansatz betriebswirtschaftlicher Gründungsprobleme. Frankfurt am Main.

Marketing in New Technology-Based Firms

Franz Pleschak, Henning Werner, Udo Wupperfeld

1.   Introducing the Subject

Together with financing, marketing constitutes one of the most difficult management tasks in new technology-based firms. Founders have hardly ever had any previous experience in this area. To begin with they are more researchers than entrepreneurs, and they fail to take sufficient account in their planning of the time and
expense hat will be incurred up to the successful market entry of the new products
and processes. The high complexity of research and development (R&D) projects
often leads to the extention of the originally planned development time, thus delay-
ing market entry. Experience has shown that founders often commence their marketing activities too late. They underestimate how long it will take to bring customers
to the decision to purchase, to counter resistance to market entry, to build up a dis-
tinctive image for their firm and to create an effective sales organization.

The complex nature of market tasks in new technology-based firms is expressed in
the following factors:

The majority of firms wish to sell their products on the investment goods market.
This market is by no means simple, because several different departments and persons are usually involved in the decision-making processes of potential customers.
The new products or processes have to be appropriate for the customers' innovation
processes and they have to be available at the right point in time. This requires very
close contact with customers.

The enterprises mainly target international markets; the German market may be
their initial market. Since new technology-based firms do not yet possess the relevant market knowledge, they have to bring in sales partners.

The innovative products and services are often systemic in character, implying the
necessity not only to sell products, but also to take on tasks of consulting, qualifica-

tion, servicing, quality assurance and maintenance. This makes great demands on the firm's potential.

New technology-based firms derive competitive advantages from their ability to be customer-related and to gain a lead in quality and technology. The three issues that need to be examined in detail are: whether the potential market is receptive, how a time advantage can be maintained, and how the market needs to be prepared for the introduction of the new products and processes.

These specific characteristics of new technology-based firms emerge clearly from the research conducted by the "Innovation Services and Regional Development" Department of the Fraunhofer Institute for Systems and Innovation Research in the scientific monitoring of three German government schemes run by the Ministry of Education, Science, Research and Technology (BMBF). These were the pilot schemes "Promotion of New Technology-Based Firms" (TOU) in the old and the new federal states, and the pilot scheme "Business Investment Capital for New Technology-Based Firms" (BJTU) (cf. Kulicke 1993; Pleschak et al. 1994; Pleschak et al. 1995; Kulicke/Wupperfeld 1996; Baier/Pleschak 1996). These enable conclusions to be drawn about the marketing concepts that new technology-based firms need and about the marketing tasks associated with the different phases in the life of an enterprise. Both points are dealt with in this paper.

2.   Marketing Guidelines for New Technology-Based Firms

At every stage in the life of a technology-based firm, there are marketing tasks to be done. In the initiation phase, market requirements and market options form the basis for establishing the business aims of the enterprise and elaborating business concepts. The "checklists" compiled for R&D projects are based on customer requirements, the competitive situation and the development of the market. As well as technical, economic, time scheduling and organizational goals they have to include the market goals. At all stages in R&D, the development results are evaluated for attainment of these targets. In parallel to the R&D, market preparations take place.

At the market entry of the new products or processes, it will become apparent whether the marketing ideas that formed the basis of the foundation and the R&D were realistic or not. Figure 1 shows a summary of the planning areas involved in marketing.

Figure 1:   Areas of marketing

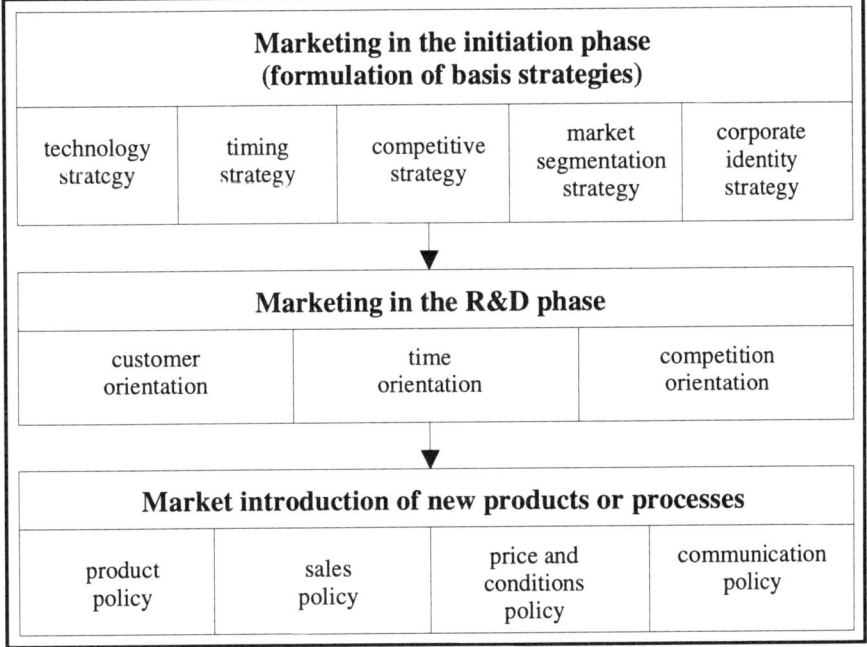

All marketing tasks are oriented towards the following guidelines:

Marketing is a task that involves the whole firm: marketing should be understood as market-oriented management relating to all functions and departments of the enterprise.

Customer orientation of the firm as a whole: the customer, and usefulness to the customer, must be the centrepoint of all considerations.

Competitive orientation: even when they are offering innovative products, new technology-based firms have to compete with other options or substitute products. Thus marketing has to be competitor-oriented in the sense that it is directed towards gaining competitive advantages, defending the firm's market position and enabling it to expand further. A firm possesses comparative competitive advantages if, from the customers' viewpoint, it has a higher problem-solving potential than its competitors.

Long-term strategic thought and action: marketing cannot be based on a specific current situation only; it requires a long-term strategic approach to all business issues.

Innovation and technology orientation: marketing is based on continual changes in the market and in technological development. Innovations are therefore an essential prerequisite for new technology-based firms in order to gain lasting competitive advantages.

Growth and profit orientation: new technology-based firms can only establish themselves permanently on the market if they can assemble the finance resources necessary to generate and market innovations in the long term, and if they have an adequate presence in their market segment. This requires a certain minimum firm size and level of turnover. Thus growth orientation and profit orientation are central components of a marketing concept for new technology-based firms.

Marketing strategies are fundamentally important for the success of the firm. This is particularly true of new technology-based firms, which have to steer their limited potential very judiciously towards the most suitable topics. Since marketing strategies represent the long-term basic orientation of the firm and require a lot of the firm's resources, it is a central task for founders to develop appropriate marketing strategies during the initiation process.

It is clear from the example of new technology-based firms in East Germany shown in Table 10 that many technology-based firms begin their marketing activities too late. This can create risks in the marketing of the new products. It is important to define marketing strategies and fix marketing milestones already in the conceptual phase of the enterprise.

Table 10:   Time of commencement of market preparations

Commencement of market preparations	Proportion, in percent (n=46)
Before founding the firm	9
In the business concept and R&D project planning phase	39
During the development process	52

Source: Bräunling et al. 1994:33

3.   Marketing During the Initiation Phase

For a business concept to have a sound and realistic basis, an information input is required about the market, about potential customers and competitors, about available resources and capabilities, and about relevant background factors in the environment of the new firm. In order to adequately meet the needs of new technology-based firms, **market research** thus has to answer the following questions:

- Who are the customers for the planned innovation, and where are they ?

- What requirements does the innovative solution have to satisfy?

- What competitors are in the field, and how competitive is their performance?

- How can the state of technical development be assessed, and

- What factors influence the firm's environment?

The necessary information can be gathered in primary research, from monitoring, questionnaires, experiments or tests. However, it is generally cheaper to make use of data in literature or in data bases (secondary research).

When elaborating the business concept, the following **marketing strategies** have to be developed: the technology strategy, the innovation and timing strategy, the competitive strategy, the strategy of market segmentation and the "corporate identity" strategy.

Technological success potentials are linked mainly with the **strategy of a techno-logical leadership**. Understood as the strategy of building up and maintaining a leading position in the development and application of product or process technologies. By achieving technological leadership position, firms create opportunities for product differentiation and the reduction of manufacturing costs. A technological leadership position may also bring pioneer profits. A leadership position in technology can be pursued with varying degrees of intensity. An active leadership strategy is particularly appropriate if the relevant technology has great importance for competition as well as a high development potential (Specht/Zörgiebel 1985). For the realization of an active technological leadership strategy, the firm must fulfil the following prerequisites:

- a high degree of technological competence, as well as clear long-term advantages over competitors in the development of the technology concerned,

- availability of sufficient resources, e.g. in terms of finance and personnel,

- organizational conditions, such as the coordination of research, development, production, marketing and sales,

- sustained, long-term development of new products and/or processes.

An aspect that is directly linked with the technology strategy is the decision about timing the introduction of a technological solution on the market. A firm which pursues a **pioneer strategy** is the first to introduce a new technology on the market. Although this type of strategy generally opens up many chances, it is also associated with a high degree of risk. The pioneer strategy is a good option for new technology-based firms particularly if they have high competence in the relevant area of technology and if they have a definite time advantage over their competitors which will enable them to realize clear competitive advantages.

The early follower enters the market shortly after the pioneer. At this point, the market is still developing and the market positions are thus not yet "cemented". However, in his market entry the "early follower" has to take account of the pioneer's activities. Thus this strategy has a strong competitive orientation. However, it can also be advantageous to delay market entry until another supplier has already gathered initial experience of the new market, since this can reduce risk. Like the pioneer, the early follower must reckon with the appearance of other suppliers on the market later.

One **competitive strategy** that is typical of new technology-based firms is **differentiation**. In this type of strategy, competitive advantages are derived from exceptional product advantages with relevance for the market (e.g. a new order of performance or reliability). A differentiation strategy can bring the advantages of high profit, a market position that is relatively sheltered from competitors and a lower sensitivity of customers to the price factor. However, the prerequisites for following this type of strategy are: a high level of R&D, fast market introduction of the innovation, effective quality management, very close contact to the customer, judicious use of marketing instruments and a positive firm "image". When pursuing the differentiation strategy, it must be taken into account that for most products a "standard quality" has emerged; without which products will be virtually unsaleable. This also has the consequence that it is growing more and more difficult for firms to differentiate their products, from the customer's viewpoint, from those of their competitors. The best chances for a differentiation strategy are generally to be found in the following directions:

• fulfilling specific customer wishes and conditions of use for the product,

• offering systemic solutions to complex problems,

• supplying top-quality services,

• developing environmentally acceptable products.

A **concentration strategy** can be advantageous for new technology-based firms aiming at market niches. Here, small firms can gain advantages over large enterprises by their greater flexibility and customer orientation. When targeting market niches, they may draw competitive advantages either from the superior performance of their products (differentiation) or from lower prices. The firm should also examine whether it should target several market niches, insofar as its potential allows. This reduces the market risk (spreading it over several products) and may result in synergy effects. Table 11 shows that the majority of firms receiving support under the TOU-NBL pilot scheme have other products or services in their programme apart from the supported R&D project.

Table 11: **Intended performance spectrum of firms receiving support under the TOU-NBL pilot scheme**

Performance spectrum	Share in % (n=212)
Marketing of results of supported R&D project only	21.2
Marketing of results of supported development, plus other products	36.3
Marketing of results of supported development, plus other services	42.5

Source: Pleschak/Rangnow 1995:24

New technology-based firms do not generally possess the prerequisites for a comprehensive cost leadership strategy.

Table 12 shows the relative frequency of adoption of the various competitive strategies, as used by 118 new technology-based firms receiving support under the BMBF's pilot scheme "Business Investment Capital for New Technology-Based Firms".

Table 12: **Frequency of competitive strategies followed by new technology-based firms (multiple entries possible)**

Type of competitive strategy	Frequency in % (n=118)
Market niche	61
Systemic solution	52
Product adaptation to customer wishes	51
Technology leader	47
Cost leader	11

Source: Kulicke 1996:149

Due to their limited resources, new technology-based firms may not be able to cover the whole market. Moreover, customer orientation is only possible if specific target groups can be defined. It is therefore necessary for these firms to develop a **market segmentation** strategy at an early stage. This includes:

- splitting up a relatively non-homogeneous total market into market segments that are as homogeneous as possible,

- selecting one or several suitable segments as target groups, and

- processing each segment in accordance with the target group involved (e.g. positioning the new technology-based firm and its products; developing segment-specific use components of the innovation, etc.).

The first stage is a **macro-segmentation**, based on the characteristics of potential customers. The following factors, for instance, could be considered as macro-determinants influencing purchasing behaviour: industrial sectors or fields of technology, firm size, organizational structure, location of customer, buying situation, innovative behaviour.

The second stage is **micro-segmentation**, in which the target groups are more narrowly defined. Micro-segmentation is a very lengthy and complex process, since firm-internal criteria have to be brought in, such as:

- the role of the decision-maker and the persons influencing the decision to buy (e.g. supporting or resisting the purchasing decision, preparing the decision, submitting expertises, etc.) (Webster/Wind 1972; Witte 1976),

- the position of these people in the firm's hierarchy,

- the technical competence of participants in the procurement process,

- attitudes to new technology-based firms, image-related reactions and cooperative behaviour.

In order to counter image-related disadvantages from the outset, new technology-based firms should have their own **"corporate identity" strategy**. This type of strategy has the following aims (Birkigt et al. 1989):

Externally, its purpose is to build up a positive image for customers, investors, co-operating partners, potential employees, etc.. This is accomplished by putting across the competence of the enterprise and the advantages of the innovation in a persuasive and credible way (image-related aims).

Internally, a corporate identity is intended to convey to employees a corporate philosophy, including marketing guidelines. This is very important, in the sense that a customer and innovation orientation can only be achieved if all employees identify with the aims of the new technology-based firm and pursue them themselves (management-related aims).

The corporate identity strategy is based on corporate culture. This is an expression of the values, traditions and ways of thought prevalent in the enterprise. At the same time, it aims to develop and disseminate corporate guidelines, thus creating a unified, consistent and positive image of the enterprise and its total offer as a whole. To achieve this, the following "corporate identity" instruments are used: visual management elements (logo, design, packaging), communication measures (advertising, public relations, sales promotions) and behaviour characteristics with customers, suppliers and cooperating partners.

4. Marketing in Research and Development

For small technology-based firms, customer orientation is an important success determinant for R&D. Taking customer needs as a guideline for the firm's activities and integrating usefulness to customers into the firm's technology policy is thus a basic principle of successful **marketing** (Töpfer/Vetter 1991). To accomplish this, customer problems that arise have to be identified, and emerging customer needs have to be recognized. The latter aspect gives impulses for the firm's innovation strategy and research tasks in the long and medium term. All experiences indicate that putting too much time and too many resources into R&D leads to "exaggerated" technical solutions that are neither recognized nor rewarded by customers.

When **interacting** with customers, they have to be persuaded of the advantages of the new technical solutions and their usefulness. The more the solution coincides with the customer's aims, the more lasting the impact of persuasion will be. The spectrum of customers and the conditions of use for new technological solutions are usually highly differentiated; at the same time, not all customers are equally important for technology-based firms. Thus it is necessary to identify the **key customers**. These are: customers who will be entertaining particularly intensive business relations with the technology-based firm in the long term; customers with an open attitude to innovative developments; customers whose names are respected on the market and can be used as references; and customers who play a progressive role in forming opinion, for instance in associations, chambers and interest groups.

If a technology-based firm has several product groups, it details **"key account"**
managers to deal with the various key customers. Their task is to put across to
customers the corporate identity of the firm, its innovative capability and its high
performance, while at the same time transferring into the firm the developments,
requirements and trends that are emerging on the market.

Customer awareness in R&D also includes keeping informed about **customer ac-**
ceptance of a new technical solution (adoption). Customer who enjoy experiment-
ing will be prepared to introduce novelties. This customer-related information spurs
and strengthens adoption processes. Information is particularly important in situations where there is a danger that customers will not have an overall view of complex technical interrelations, or in sensitive areas of technology where an attitude of
confidence and understanding of the innovation first has to be built up (cf. Nieschlag et al. 1991).

Customer awareness in R&D also implies

- identifying constant and variable customer requirements, and investigating the
  influence of the extent of their fulfilment on the decision to buy,

- analysing customer behaviour, if potential customers would be giving up traditional or established supply or cooperation relations,

- recognizing pressures and motives that may lead customers towards reorganization and innovation,

- awareness of background factors that influence the purchasing decisions of customers (e.g. the influence of associations, societies, public committees, laws and
  standards),

- taking account of regional differences in customer behaviour that may arise, for
  instance, from consciousness of traditions, conventional attitudes or an enthusi-
  asm for experiments,

- examining customers' ability to finance future orders (financial standing and creditworthiness).

The **duration of the process** is an important success determinant. Among the conditions pertaining to technical innovations, particular attention should be paid to
"time to market" as a determinant in achieving a high market share and asking high

prices. If innovative products or processes are made available to them early on, customers gain economic advantages by expanding their capacity, increasing their flexibility, conserving resources, gaining a lead in experience and enhancing their image. Customers are exposed to risks if the new solutions are not yet technologically mature at market entry. It is a prerequisite for acceptance of the "time lead" situation by customers in industry - and consequently for the possibility of charging higher prices - that the customer consciously exploits the time advantage, and that the advantage has a lasting impact. The aim of time management is to establish the optimal timing for market entry and, based on this, to schedule the milestones in the innovation process.

A time orientation in R&D implies considering the whole innovation process in its complexity when planning the **time schedule**. The more complex the innovation, the more components the schedule will have. If individual aspects of the innovation process are neglected, the time scheduling of projects will very soon become unrealistic.

When fixing the R&D goals and evaluating the interim and final results of the R&D process, it is decisive for success on the market not only to consider customer aspects and timing elements, the internal factors of the firm's potentials, scope of manufacturing, organization and viability, but also the **behaviour and development strategies of competitors**. Only if the R&D schedule and the R&D results can build up competitive advantages for the firm it will be possible to establish the new technical solution in the market.

Competitive analyses, customer analyses, market analyses form the basis for the **concept of the firm's competitive advantages** and their definition in the checklist. Methods that support this concept are: analysis of strengths and weaknesses, analysis of chances and risks, benchmarking, evaluative methods such as assigning points and weightings, firm-related and customer-related feasibility calculations. R&D projects are based equally on the possibilities arising from technological development and on pressure from market forces. Scientists, research workers, designers and managers in the firm itself may be a source of ideas for new products and processes, as well as may customers (through direct questioning, tests, group discussions, suggestion systems), competitors (through market observation, comparison of levels, visits), patent attorneys, business acquaintances, suppliers or consultants.

Faced with fragmented markets, a scarcity of capital, high development costs and short product life cycles, firms must examine whether the targeted customers will welcome the new product and recognize its advantages, and what cost-benefit relationship will make them willing to buy. A high level of innovation may meet with open attitudes in some target groups, but others may react with scepsis. From the competitive situation and the customer situation, the permissible market prices and targeted costs for the checklists are derived. Methods used in this process are calculations of the benefit to customers, price comparisons, value analyses, identification of target costs and the comparison of variants.

From the comparison with other firms, **barriers to market entry** emerge for the firm, for instance because competitors

- have economies of scale in manufacturing and marketing,
- have stable customer relations with high market shares and traditional customer structures, and have tied their customers down,
- are blocking the firm's access to sales channels,
- have a lead in R&D know-how,
- have patent rights,
- were the first to be successful on the market, and the costs of opening up the market for "followers" are too high,

or because the firm itself is not in a position to

- ensure compatibility of the new products,
- get together the capital needed for production build-up and market introduction,
- provide references,
- offer lower prices or higher quality than its competitors,
- satisfy authorisation or licensing requirements for the new product.

These barriers can be actively countered by measures which are a part of the R&D checklist. If the measures for overcoming the barriers to market entry prove to be unrealizable, then it would be better not to undertake the R&D project.

Pilot marketing is used to carry out initial testing with customers in order to accumulate user experience, test the functioning of the technical solution and draw conclusions for perfecting it. Customer testing is preceded by laboratory tests performed by the technology-based firm. However, since these are not based on the concrete conditions of use and application, reliable predictions about the functioning of the product cannot generally be made on the basis of laboratory tests. Moreover, the transition from laboratory tests to practical testing is often associated with increasing the dimensions of the product, which may bring technical risks.

After the technical solution has been revised and perfected, it is necessary from the point of view of marketing to acquire **lead users** who will use reference products or, in the case of processes, will construct reference options. Reference options have the effect of promoting sales, if the key customers have a reputation as innovative, efficient and competitive enterprises, and if they are prepared, together with the technology-based firm, to bring the advantages of the innovation to the notice of other customers. For the technology-based firm, this marketing strategy necessitates intensive advice to reference customers, and constant service to ensure that the reference option remains in full working order. Since innovations need to be explained to customers, lead users have to be initiated into the technical know-how of the technology-based firm. They can be economically motivated to support the acquisition of order by the firm.

Successful tests and positive user experience by "reference" customers are a prerequisite for informing the targeted breadth of market about the innovation (broad-based marketing), triggering a second wave of publicity for the innovation. This strategy is intended to make the advantages of the new technical solution apparent to the targeted customers, to demonstrate that the innovation has been tried and tested, can be introduced into the customer's existing structures (compatibility) and that the technology-based firm will supply all the services necessary to make the innovation useful to the customer (complexity). At this stage, the "corporate identity" strategy of technology-based firms is directed towards building up customers' confidence in the innovation. This will create an image for the firm.

5.   The Market Introduction of New Products and Processes

Product naming, packaging and design are important cornerstones when laying the foundations for the market introduction of new products. With technology-based products, the basic function is enhanced by the addition of customer service options such as guarantees, servicing and maintenance and services supporting introduction and use. The "service mix" may emerge as a decisive differentiating characteristic, and the increased usefulness to the customer that results from the combination of product performance and services may constitute a decisive competitive advantage. In this context, it is necessary to define those characteristics which particularly distinguish the product from the products of competitors ("unique selling propositions") and are central to the purchasing argumentation (cf. Bruhn 1990; Töpfer 1991).

Guarantees reduce the risk incurred by the purchaser of the product. Guarantees that exceed the legal requirement generate feelings of confidence and enhance the firm's image. Particularly in the market introduction phase, great attention should be given to ensuring very high quality supply.

The tasks of **customer service** increase with the complexity of the product. These include installation, inspection, servicing, repairs, providing spare parts, but also instructing customers and dealing with queries and problems. The feedback of information from customers to the firm makes the manufacturer aware of existing problems and generates ideas for new solutions. As well as customer satisfaction, the aim for the technology-based firm is to identify latent or future problems that customers may have, and exploit this information itself.

Support services include the supply of services such as training for product users, the provision of training documentation and technical consulting on applications. The smooth introduction of new technologies is often crossed by the resistance or qualification deficits of employees in the firm where it is being introduced. In order to avoid this situation, employees affected by the introduction of the new technology should be informed in good time and should receive training if necessary (Unterkofler 1989).

Three **sales channels** are open to new technology-based firms:

- direct marketing,

- indirect marketing and

- a combination of direct and indirect marketing.

Direct marketing is characterized by close contact between the manufacturer and the customer. People who are directly involved in the development process are better equipped to meet the need for explanations arising from the high-tech character of the products than business agents or sales intermediaries. Experience has shown that it is advantageous in the introduction phase to choose direct marketing. The firm's own personnel has more know-how and is more strongly motivated in selling situations. There is a more direct flow of information and this remains accurate, especially if sales and development work are carried out by the same personnel. The direct feedback of information from customers about problems in using the product and their further wishes for solutions should be consciously cultivated and methodically supported.

Indirect marketing lessens the burden on the personnel resources of the firm and under some circumstances may offer the advantage of established, efficient sales channels and abundant marketing experience. However, sales intermediaries (retail, wholesale, commercial agents) are generally interested in a whole range of products, so that it is mainly standardized products that are sold via these channels.

It is essential for new technology-based firms to take on some of the sales activities themselves, in order to cultivate customer contacts and be constantly informed about new problems. A typical set-up is the combination of direct and indirect selling, in which new technology-based firms limit their selling activities to processing the most attractive market segments and bring in sales intermediaries for the rest of the market processing. In this way, the advantages of both basic strategies can be combined. Table 13 shows the shares of the individual sales strategies in the sales concepts of new technology-based firms.

Table 13:   **Sales concept of new technology-based firms (share in %)**

Type of sales concept envisaged	Old federal states (n=93 enterprises)	New federal states (n=212 enterprises)*
Selling by own firm only	22	26
Selling by own firm and cooperation partner (multiple entries possible) of which	59	67
Selling with support from East German enterprise		13
Selling with support from West German enterprise		49
Selling with support from Western business partner		19
Selling by partner only	10	1
No statement as yet	9	6

* multiple entries possible

Source: Kulicke 1993:36; Pleschak/Rangnow 1995:39

As a prerequisite for sales decisions, the following aspect have to be investigated:

- the influence of sales cooperation on market segments, possible quantities and the pace of market entry,

- comparison of the capital required to open up the market using ones own sales organization exclusively, or complementing this with cooperation partners,

- the consequences of granting sole rights to sell,

- the relevance of turnover from sales cooperation both for ones own firm and for the cooperating firm,

- alternative strategies if a sales partner drops out,

- qualification of the sales partner, which should enable him to explain the products to customers and act as their technical contact partner, as well as establishing durable customer-supplier relations,

- the danger of technical know-how leaks via the sales partner.

Price and conditions policy includes all measures involved in fixing conditions for the firm's performance. As well as price structuring, they include all decisions about discount policy, payment and delivery conditions and credit policy. The **price structuring procedure** for a new product or process may be cost-oriented, demand-oriented or competition-oriented (Kotler 1989).

In **cost-oriented price structuring**, a profit margin is usually added to the cost price ("cost plus"). In this way it is easy to work out the minimum feasible price level. Purely cost-oriented price levels are only customary in situations where comparisons and alternatives do not exist because there are no equivalent products. The method of **demand-oriented price structuring**, on the other hand, is based not on costs but on customers' expectations of usefulness and the intensity of demand. The main orientation parameter in **competitively-oriented price structuring** is the prices of competitors. **Integrated price structuring based on cost-benefit analysis** represents a combination of these three basic strategies. Among the possible price strategies for new technology-based firms, skimming strategy and price structuring according to the average prices of competitors carry more weight than the market penetration strategy.

For new technology-based firms, the greatest problem associated with market entry is their lack of **image**. Potential customers not only have to assess the performance of the innovation being offered; they also have to assess the survival chances of the young technology-based firm, because the services agreed on at the time of sale (guarantee, servicing, maintenance, delivery of spare parts, etc.) are only assured provided the NTBF survives. Thus **communication policy** has to strike a balance between general public relations and product-related advertising.

The aim of **public relations** it to enhance the firm's reputation, making it better known and building up a positive image so that it is regarded as a reliable and competent business partner, thus generating a positive attitude to the firm. It is important to cultivate a professional image, e.g. by the design of pamphlets and standsat fairs and to convince customers by a professional manner.

The aim of **product-related advertising** is to disseminate information about the firm's performance programme which will influence the buying decision. According to the results of research on diffusion, there are five characteristics that support the diffusion of innovations in the market place, if due consideration is given to them in marketing activities. The acceptance of new products on the market is facilitated if the advertising message emphasizes (cf. Geschka 1990)

- the **relative advantage** of the products,
- their **compatibility** with the user's previous experience,

- the **simplicity** of the application,

- the **low risk** for the purchaser and

- the fact that the product can be **demonstrated**.

Fairs and exhibitions constitute the central communication instrument for new technology-based firms. Because of the great importance of this instrument, a great deal of care and attention has to be given to the preparation and follow-up activities for trade fairs. During preparation, it is important to inform customers of the firms' presence at the fair through press advertisements or sales letters, and invite them to a personal discussion. A written record should be kept of the results of the sales talks and of the customer problems mentioned in this context, so that they are available for analysis later.

Publications in specialist journals and presence at scientific events as advertising measures are also very specifically directed towards certain target groups. Particularly in the first two phases of the adoption process, the awareness phase and the interest phase, information is absorbed by the customer mainly from mass media, whereas in the last two phases, the experimental phase and the purchasing phase, personal contacts serve as the main source of information.

6. Summary

This chapter gave an idea of the content of marketing concepts for new technology-based firms and highlighted the points that need particular attention during the expansion phase. Marketing tasks change, depending on the stage in the life cycle of the enterprise. In the initiation phase, the basic strategies and the overall approach are established. The decision is made as to whether the new technology-based firm will aim at a technological leadership position or adopt a position as a technology follower. The timing strategy fixes the time of market introduction. Competitive strategy covers decisions on how to retain comparative advantages over competitors in the long term. Market segmentation splits up the total market into relatively homogeneous groups and selects the most attractive segments. A sound, reliable segmentation is highly significant for the market success of new technology-based firms since, due to their limited resources, they are unable to process the whole market. Market segmentation clarifies the prioritizing of customer groups. Corporate identity strategy aims to create a positive external image for the firm and to transmit a corporate philosophy internally to employees.

In the R&D phase, the technical realization of the new product or process occupies the foreground. Customer orientation, time orientation and competitive orientation are eminently important at this stage in order to avoid the dangers of development by-passing customers' needs, loss of profit due to long development times, or competitors catching up.

The market introduction phase is decisive for the economic success of the innovation. The marketing instruments at this stage are tools designed to shape and influence the market. The harmonization and coordination of the instruments used is important here.

Marketing emphasis varies from one technology-based firm to another. The choice of marketing measures and instruments may be influenced by the following factors: the novelty and complexity of the product, the systemic nature of the innovation, the area of technology concerned, the degree of high-tech involved, the market and market segments targeted, the competitive situation, customer structure, customer behaviour and the demand situation. Each of these influences may be present to a varying degree. It is thus an important and creative management task for young firms to find the marketing measures and instruments most suited to their purpose, and adapt them to meet their individual needs.

7. Bibliography

Baier, W./Pleschak, F. (Eds.) (1996): Marketing und Finanzierung junger Technologieunternehmen. Wiesbaden.

Birkigt, K./Stadler, M./Funk, H.J. (Eds.) (1989): Corporate Identity. 4th Edition. Landsberg a. Lech.

Bräunling, G./Pleschak, F./Sabisch, H. (1994): Chancen und Risiken von im Modellversuch TOU-NBL geförderten jungen Technologieunternehmen. 3. Analysebericht. ISI: Karlsruhe, Dresden.

Bruhn, M. (1990): Marketing. Wiesbaden.

Geschka, H. (1990): Marketing-Konzeption für neue Produkte. In: Poth (Ed.): Marketing. Düsseldorf.

Kotler, P. (1989): Marketing-Management. Stuttgart.

Kulicke, M. u.a. (1993): Chancen und Risiken junger Technologieunternehmen - Ergebnisse des Modellversuchs "Förderung technologieorientierter Unternehmensgründungen". Heidelberg.

Kulicke, M./Wupperfeld, U. (1996): Beteiligungskapital für junge Technologieunternehmen - Ergebnisse des Modellversuchs "Beteiligungskapital für junge Technologieunternehmen" (BJTU). Heidelberg.

Nieschlag, R./Dichtl, E./Hörschgen, H. (1991): Marketing. 16th Edition. Berlin.

Pleschak, F./Rangnow, R. (1995): Ergebnisse des BMBF-Modellversuchs "Technologieorientierte Unternehmensgründungen in den neuen Bundesländern" der Jahre 1990 bis 1994. 7. Analysebericht. ISI: Karlsruhe, Freiberg.

Pleschak, F./Sabisch, H./Wupperfeld, U. (1994): Innovationsorientierte kleine Unternehmen. Wiesbaden.

Pleschak, F./Werner, H./Wupperfeld, U. (1995): Marktbewährung geförderter junger Technologieunternehmen. 8. Analysebericht. ISI: Karlsruhe, Freiberg.

Specht, G./Zörgiebel, W.W. (1985): Technologieorientierte Wettbewerbsstrategien. In: Marketing ZFP (7), 161-172.

Töpfer, A. (1991): Marketing für Start-up Geschäfte mit Technologieprodukten. In: Töpfer, A./Sommerlatte, T. (Eds.): Technologiemarketing. Landsberg a. Lech.

Töpfer, A./Vetter, H. (1991): Anforderungen an das Technologie-Marketing in mittelständischen Unternehmen. In: Töpfer, A./Sommerlatte, T. (Eds.): Technologiemarketing. Landsberg a. Lech.

Unterkofler, G. (1989): Erfolgsfaktoren innovativer Unternehmensgründungen. Ein gestaltungsorientierter Lösungsansatz betriebswirtschaftlicher Gründungsprobleme. Frankfurt am Main.

Webster, F.E. jr./Wind, Y. (1972): Organizational Buying Behaviour. Englewood Cliffs.

Witte, E. (1976): Kraft und Gegenkraft im Entscheidungsprozeß. In: ZfB (46), 319-326.

Crises of New Technology-Based Firms

Joachim Hemer

1.   Introducing the Problem

Building up a new technology-based firm (NTBF) to the point where it becomes a competitive enterprise with a stable position in the market makes great demands on the founders. Although founders primarily contribute technical know-how, they often do not possess sufficient business skills and the owned capital that they are able to bring into the new foundation is usually small. Their state of information regarding the market and legal and financial questions is often only limited. They are thus highly dependent on cooperation with external investors and providers of information, and have to acquire new abilities and additional know-how during the initial business years.

The creation of a new business unit which has to be embedded into national and international sales and supply markets is generally punctuated by a series of crises, both large and small, before the newly-founded firm reaches a state of equilibrium. Previous studies by the Fraunhofer Institute for Systems and Innovation Research (ISI) on new technology-based firms in Germany (cf. Kulicke et al. 1993 and the ISI reports referred to there) have provided a wealth of knowledge about the specific problems of NTBFs, their growth patterns and their needs for consulting and financing. The following barriers emerge as typical of technology-based firm foundations:

- there is a lack of capital, particularly too little owned capital,

- high demands are associated with build-up and management and with the development and marketing of an innovative offer,

- founders have know-how and experience deficits in all aspects except technical issues,

- barriers to market entry hinder a smooth, rapid market entry,

- there is a lack of qualified personnel, particularly in marketing and sales.

It was the complete - or near-complete - failure of firms of this kind that provided the starting-point for previous investigations of the failure rates of public-supported NTBFs, and their causes (e.g. Kulicke 1990: 18ff.; Kulicke et al. 1991: 28ff.; Wupperfeld 1993; Kulicke 1994). These studies centred on identifying the constellation of negative factors leading to failure. The present contribution, on the other hand, concentrates on the **causes and effects of crises** that typically occur in the start-up process and the subsequent business path of NTBFs, but do not necessarily lead to failure.

In the following chapter, many experience deficits and gaps in the know-how of NTBF founders are named as reasons for problems or crises. The reference picture of an "ideal" founder or entrepreneur on which these reflections are based is certainly rarely encountered in real life, even in the management of established small and medium-sized firms. The latter, in any case, are not faced with the demanding tasks of starting-up a completely new business from scratch; moreover, they are able to look back on long years of business experience - unlike the founders of NTBFs, who first have to accumulate the necessary experience under difficult conditions.

The explanatory models for crises discussed in the literature (see below) are only applicable to NTBFs to a limited extent, since they relate to established firms whose infrastructure and organizational structure have already been built up, and to later phases of development such as market expansion, extending the sales apparatus or expanding production. By contrast, this chapter is concerned with the characteristics, causes and courses of crises in the **early development phases of NTBFs**. It is empirically based on in-depth interviewing of 42 German NTBFs in 1993/94; these were all firms which, at least two years before the interview, had received subsidized investment capital under the BJTU pilot scheme ("Business Investment Capital for New Technology-Based Firms" run by the Federal Ministry of Research and Technology, which was accessible from 1989 to 1994).

Most of these NTBFs had already been in existence for some years before seeking investors. The initial phase of business development was financed from a variety of sources (personal resources, income from a service business, public promotion programmes and bank credits, usually in combination). In 55 percent of the NTBFs interviewed, the decision to seek investment capital had not been a strategic decision taken at the time of foundation, but became necessary in many cases to combat a financing bottleneck or, less frequently, when the existence of the firm was acutely endangered. The main reason for taking investors on board was an additional need for capital.

The survey of the relevance, causes and paths of crises in the 42 NTBFs is not confined to the period following the injection of capital, but also covers the whole course of the businesses from their point of formal foundation onwards. For many NTBFs undergoing crises, the injection of investment capital constituted an opportunity for strategic re-orientation of the enterprise, upgrading of R&D or marketing activities and stronger development of the firm's innovation and growth potential. The observation period of this survey falls mainly within the German economy's biggest recession since the war. As a consequence, the problems confronting practically all newly-founded firms in their first years of existence were substantially magnified. It can thus be assumed that the NTBFs in the survey underwent more crises than would be typical under more favourable economic circumstances.

The **main research questions addressed** by the study on which this chapter is based (cf. Hemer/Kulicke 1995) were the following:

- What are the signs of crises in NTBFs and how can they be recognized early on?

- Do the crises courses of NTBFs follow specific patterns ?

- Can typical cause-and-effect chains be identified for crises?

- How can the crises of NTBFs be classified?

- Can "hands-on"servicing by investors have a positive influence on the course of the crisis or on its avoidance?

- What measures do investors resort to in order to avoid and combat crises?

2.   Recognizing a Crisis in a NTBF

2.1   Definition, Causes and Effects of Crises

In the numerous works addressing insolvency and crisis research, no uniform definition of the term "business crisis" has been arrived at (cf. bibliographical indications); however, there are approaches which attempt to describe crises and distinguish them from business failure (Klein-Blenkers 1976: 5ff.; Gerybadze/Kulicke 1990: 6; Schimke/Töpfer 1986: 10f.; Marré 1986: 62; Paul 1985: 246).

The risks besetting new technology-based firms are particularly numerous and great (cf. the detailed account in Kulicke et al. 1993); any one of them may theoretically trigger a crisis. It would be inappropriate, however, to designate every deviation from a target or disturbance in equilibrium as a crisis. Following Gutenberg (1989), the following definition is used: **an NTBF is in a state of crisis if its existence is seriously endangered by an unfavourable constellation of disruptive parameters.** In NTBFs, danger of this kind can appear over a relatively short time horizon (e.g., within a year) and then disappear - in other words, a crisis is not first signalled by lasting imbalances. Crises can be overcome, remaining then only as episodes in the history of the firm. Among other things, a crisis may "only" endanger the present success of the firm ("success crisis") and not necessarily its growth potential. Thus crises can occur without dramatic consequences. However, they may also lead to fundamental re-structuring and change in the substance of the firm (for instance, from manufacturing firm to service-only enterprise). The dramatic final episode of a crisis is liquidation, i.e. the "final curtain" for the firm. This is preceded by insolvency, the strongest manifestation of a crisis. This in turn is the result of several critical phases. Paul (1985: 247) first names the liquidity crisis - triggered by a "success crisis" - as endangering the success targets of the firm (profit and turnover goals, etc.). This form of crisis is preceded, however, by so-called "strategic crises" which threaten the success potential of the firm without interfering with its day-to-day running. Departures from stability in some business variables are also regarded as crisis characteristics (Paul 1985: 245). Lastly, crises may also be attributable (at least in part) to insufficient management capacities or skills (Paul 1985: 247f.; Schimke/Töpfer 1986: 245). According to Marré (1986: 62), firms sometimes enter a time of crisis solely because their financial controlling is in a state of disorder.

The risks NTBFs are exposed to differ substantially from the risks that beset established enterprises and non-technology-based firms. The following main determinants influence the development of NTBFs (Kulicke et al. 1993: 24f.):

- the founders' target system with regard to business development,
- the spectrum of activities, business fields and performance programme of the NTBF,
- financial and non-financial resources (of founders, and third party),
- exogenous market and competition factors.

The founders themselves play a central role: the majority of founders are still inexperienced as businessmen at the time of founding, and thus only possess up to a point the experience necessary for the start-up and management of an NTBF. Juxtaposing the deficits in know-how and experience with the highly demanding task of building up a stable and competitive NTBF, it is understandable that **during preparations for founding**, problems typically arose in the following fields:

- defining the business goals and strategies,
- establishing and delimiting the main business field or market segment in which the new firm wished to compete,
- drawing up a realistic cost and finance plan for the first business year,
- carrying out demand, market and competition analyses for the products, processes or technical services to be developed.

After the formal foundation, NTBFs are confronted by particular risks both of an endogenous and exogenous nature. Table 14 gives an overview of the main risk factors and the extent to which they are observed to affect NTBFs (cf. also Albach/May-Strobl 1986). The table emphasizes the numerous parameters that come into question as causes for individual crises. Thus a monocausal or linear explanatory model is not adequate; there is a broad consensus in scientific discussion that the interrelations between the causes and effects of crises, and their mainfestations, are case-specific, complex and multicausal (in view of these complex interactions in crises, Töpfer 1990: 164ff. distinguishes "the levels of cause, effect and symptoms", levels which cannot be clearly separated). Crisis situations in one functional area of an enterprise may in fact be the result of another causal factor; they themselves may then affect other functional areas. Crises may constitute the beginning, the end or the mid-point of a chain of problems within an enterprise.

Table 14: Risk factors in NTBFs

Risk factor	Constellations observed
Product or service programme	– Shortage of resources forces limitation to a few products only (low diversification and low distribution of risks). – The initial products are generally technical innovations. If they are not yet mature at market entry, they are associated with technical risks and may require adaptation or improvement developments. – The initial products first have to gain market acceptance and do not yet guarantee an assured turnover.
Financial resources	– The capital basis is usually underestimated and a valid assessment of the need for capital is difficult because the activities to be financed are associated with imponderables and uncertainties. – Acquiring capital is a time-intensive process, especially if the requirements of various different investors have to be coordinated (the "capital procurement merry-go-round "). – The firm's lack of a track record and the lack of collateral together with the high degree of innovation of the products, make banks less willing to give credit.
Controlling/ accounting	– The necessity for an efficient accounting system and for controlling is frequently underestimated.
Personnel	– Even if the business situation is not yet stable, it is essential to engage qualified personnel on a permanent basis and build up a regular workforce. However, this strategy constitutes a substantial risk in terms of cost.
Market situation	– Every newly-founded firm lacks the image of a stable, reliable supplier of technically mature products accompanied by the relevant services and continuity of product care. – Established competitors hinder the market entry of new market participants by defensive manoeuvres. – Existing customer-supplier relations, formed over years of purchasing transactions, impede the placing of initial orders with a new competitor. – Customers often need to be convinced of the perfomance of innovative products by time-intensive demonstrations.
State regulations, certification and approvement procedures, etc.	– Unlike large, established suppliers, new, small enterprises are not in a position to exert any influence on state regulations (e.g. air pollution standards), but have to adapt to them. – State and sectoral authorisation certification and approvement procedures are time-consuming and cost-intensive; new enterprises cannot influence their course very much.
Overall or sectoral economic situation	– "Wait-and-see" buying behaviour on the part of customers in a recession greatly hinders the market entry of new suppliers. Even formally agreed orders may be cancelled at short notice.

2.2   Crisis Indicators and Crisis Symptoms

From what parameters can the (imminent) crisis, i.e. the threat to the survival of the NTBF, be recongnised? In science and in business practice, both the quantitative and qualitative criteria generally suggested as crisis indicators are often dependent on success criteria. Thus the terms success, crisis and failure are very closely related and, according to many authors, can be measured or described using the same indicators, albeit with "positive" or "negative" prefixes (Töpfer 1990; Krüger 1988).

In order for a method of identifying crises in NTBFs in the early build-up phase to be reliable, it has to include several criteria at the same time. A crisis is not an occurrence limited to one point in time; a certain time period has to be observed and, in the case of NTBFs, this may reach back to the pre-foundation phase. However, if problems or risk factors crop up this does not necessarily lead to a crisis. A crisis only occurs if an unfavourable constellation of endogenous and exogenous factors is present in the NTBF concerned. Therefore in theory crisis identification should tend to involve as many criteria as possible, although in practice this approach does not usually work out due to lack of data. Thus attempts to identify, or even predict, a crisis are always subjective to a high degree and are dependent upon the yardstick applied by the observer.

3.   Crises in New Technology-Based Firms

3.1   Empirical Basis

Between September 1993 and April 1994, personnel of the "Innovation Services and Regional Development" Department of ISI conducted personal interviews with people in leading positions (usually founders and/or managing directors) in 42 of the 48 NTBFs which, in September 1993, had already been benefiting for at least two years from the BJTU pilot scheme and were still in existence at that time. Relative to this group of firms, this represents virtually the whole sample. The minimum two-year limit was chosen under the premise that the build-up phase of the firm would already be far advanced, the R&D phase would be completed and a new, marketable product would have been brought to market, thus enabling statements to be made about the relevance and impacts of crises. Table 15 first gives an overview of important characteristics of the 42 firms interviewed.

Table 15:   Characteristics of the 42 firms interviewed

Characteristic	Frequency
Foundation year	
– up to 1985	12 %
– 1986/1987	26 %
– 1988/1989	26 %
– 1990	26 %
– 1991	8 %
– 1992	2 %
Turnover category (1993, adjusted due to missing values)	
– less than 0.5 million DM	6 %
– 0.5 to <1 million DM	31 %
– 1 to <2 million DM	31 %
– 2 to <5 million DM	16 %
– 5 to <10 million DM	7 %
– 10 to <20 million DM	6 %
– 20 million DM and over	3 %
Workforce (1993, adjusted due to missing values)	
– less than 5 employees	8 %
– 5 to 9 employees	36 %
– 10 to 19 employees	38 %
– 20 to 49 employees	15 %
– 50 or more employees	3 %
Type of investor (lead investors only)	
– seed capital companies	26 %
– venture capital companies	14 %
– societies for the promotion of small and medium industry, funds of the federal states	19 %
– business investment companies of banks and savings banks	17 %
– credit institutions	14 %
– others	10 %
Form of capital engagement	
– dormant equity	57 %
– direct participation	17 %
– conditional loan	12 %
– combination of direct participation with dormant equity or loans from shareholders	12 %
– loans from shareholders	2 %
Technological field of the supported products	
– information and communication technology	62 %
– mechanical engineering and mechatronic equipment	9 %
– process technology, machinery and plants	24 %
– others	5 %

3.2   Relevance of Crises in the Sample

The 42 founders interviewed were asked to mention problems, or even crises, which they had had to cope with in the course of the business so far. Eight of them stated that there had not been any big problems or crisis situations, but 34 mentioned serious problems since foundation. This rate should not be considered surprisingly high, as the NTBFs interviewed were largely still in the build-up phase, during which the risks specific to new technology-based firms are particularly strongly in evidence. Although serious problems were mentioned, the interviewees did not always describe them as also constituting crises. Measured by other criteria (such as non-liquidity, negative development of profits), not all the 34 NTBFs saw themselves as being in a dangerous state of crisis, either at the time of the interview or in the preceding years; rather, they felt that they were in temporary difficulties which they experienced as normal in day-to-day business.

Thus the self-assessment of the entrepreneurs does not always suffice as a valid criterion for the identification of actual crises. For this reason, ISI made its own assessment on the basis of the founder's statement, the overall impression of the interview and the data available on the NTBF, in the light of objective facts and the definition of crises proposed in Section 2.1 above. The sample of 42 NTBFs interviewed were classified as follows:

- **8 problem-free cases**: these firms stated that they had no serious problems and therefore no crises. This category also included a few NTBFs which were still in their scheduled R&D phase and therefore still had market entry ahead of them.

- **7 problem cases**: these firms perceived themselves as confronted by serious problems, but these were not grave enough to be considered as crises.

- **27 firms in crisis**: these NTBFs experienced crises which were serious or threatened the survival of the firm; however, in some cases these could be overcome.

Independently of this classification, the NTBFs were also differentiated according to their degree of success, since the occurrence of problem or crisis situations does not necessarily exclude the possibility of success later on. On the basis of data on the firm at the time of the interview and the general impression of the interview, bringing in all available information, the following (naturally subjective) evaluation of the 42 NTBFs interviewed was possible:

- **7 successful firms**: from the current standpoint these can be assessed as particularly successful, even if their build-up phase was not free of problems or crises.

- **Approx. 25 NTBFs with promising prospects**: these are set on a very promising course (at least at the time of the interview). This category includes NTBFs experiencing crises or serious problems which, according to current assessments, they have been able to overcome successfully.

- **Approx. 10 NTBFs with poor prospects**: these firms have to be regarded as poor prospects or potential failures. They include one firm which is still very young and has no substantial problems, but has not yet entered the market.

Figure 2:   Distribution of the sample into success categories and problem categories

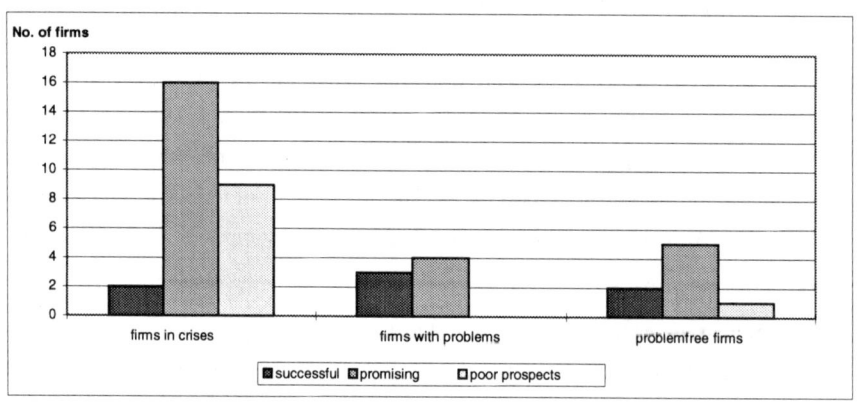

Thus, classification as an enterprise in crisis is not equivalent to failure of the firm. The course taken by the first business years of an NTBF is dependent on numerous factors and imponderables and therefore makes such stringent demands on the founder's performance that crisis situations are almost inevitable. They have to be regarded as typical occurrences in the process of building up NTBFs.

3.3 Crisis Indicators that can be Measured by Quantitative Parameters

The classical indicators for the measurement of success are mainly turnover and profit or, derived from them, the return on turnover and the return on invested capital, and the development of these parameters over time. Pursuing a consistent economic argument, it must always be the ultimate aim of the business to achieve an attractive return for partners and investors. Therefore disturbances in the development of profit or turnover are clear symptoms of crisis. In the following paragraphs, the development of turnover and profit in successful and less successful NTBFs in the sample are compared.

Figures 3 und 5 show the average **turnovers** and **annual results** (after elimination of extreme values) of firms in the crisis and non-crisis categories (only 36 of the 42 NTBFs supplied us with complete series of data on their turnover up to that time). There are no substantial differences in turnover between the two groups. On average, both groups show a continuous growth in turnover (the reduction that appears in the crisis-free category of NTBFs for the fifth business year is due solely to the fact that some firms in this group were only four years old, but growing rapidly. However, when calculating the average their turnover figures could only be included for the first four business years).

Figure 3: **Comparison of average annual turnovers (after elimination of extreme values)**

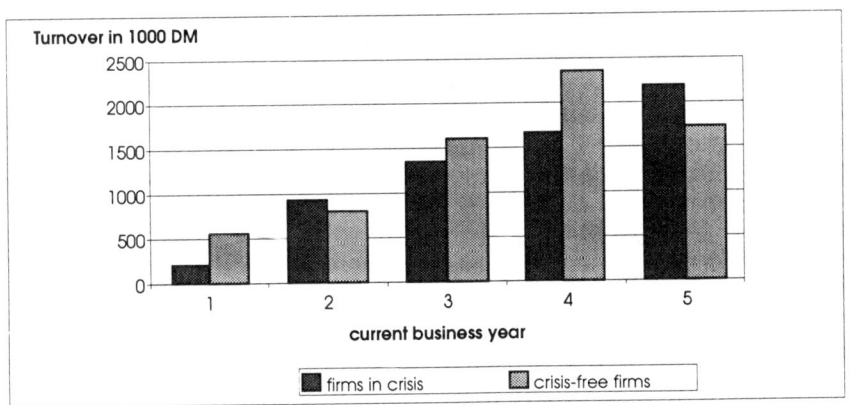

In addition to personal interviews with 42 selected NTBFs, in mid-1994 ISI addressed a written questionnaire to all new technology-based firms that had benefited from the BJTU pilot scheme, asking them about their economic development (von Wichert-Nick/Kulicke 1994: 13ff.). In this way, data were gathered on a total of 118 enterprises. When making statements on their growth, various growth stages were defined, based on the level of turnover reached, the number of years in business and the rapidity of growth. The frequency of low, medium or high growth in crisis enterprises and crisis-free NTBFs can be seen in Figure 4. The expected differences between the two groups already emerge clearly. If one examines the correlation between the incidence of crises and the rapidity of growth in the NTBFs, these differences become clearer still: in the course of their existence so far, crises have occurred in only 44 percent of the fast-growing firms but in 64 percent of NTBFs with medium growth rates, and in 78 percent of NTBFs with low rates of growth. Thus, as one would expect, crises do have an impact on the rate of growth. On the other hand, however, fast-growing NTBFs are not necessarily crisis-free.

Figure 4: **Relationship between rate of growth and incidence of crises**

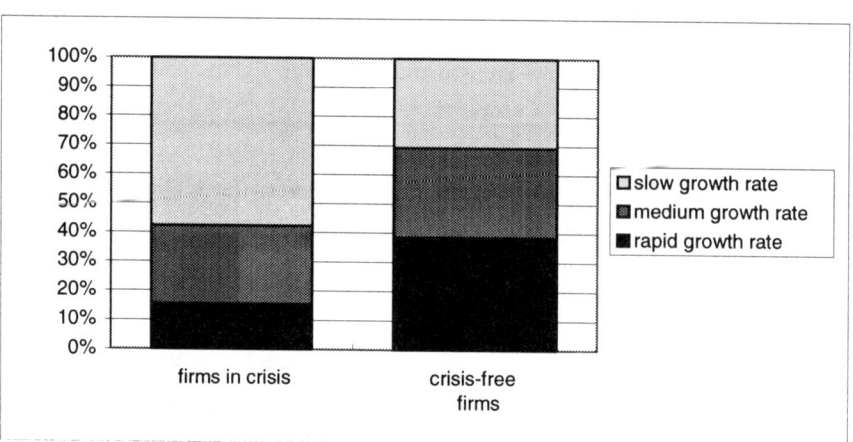

The NTBFs were generally into red figures during their first few business years. Crisis-free NTBFs, however, already reached the break-even point in the fourth business year, whereas on average neither all NTBFs as a whole, nor the crisis group, were making a profit at the time of the survey (see Figure 5). The average losses of the crisis group went up continuously, whereas the overall average losses for all firms showed a fluctuating result. The crisis-free NTBFs were quite distinct from this pattern: their profit development was already showing a continuous and - thanks to some extremely succesful NTBFs - rapidly-rising trend from the fourth year of business onwards.

Figure 5: **Comparison of the average annual results (after elimination of extreme values)**

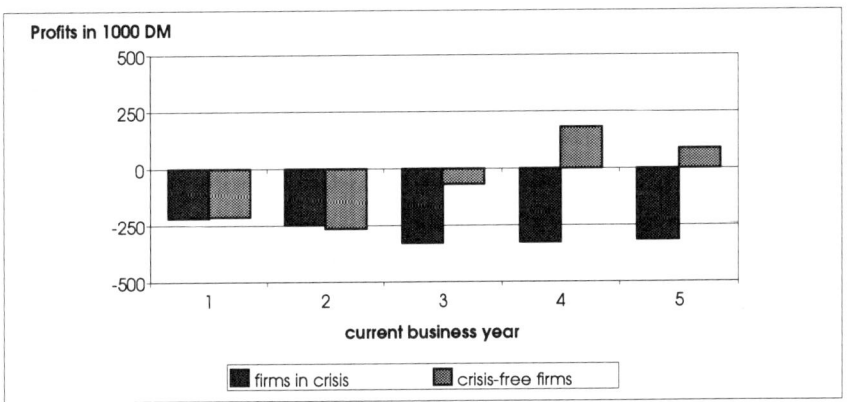

If the NTBFs are considered individually, the crisis symtoms are clearly recognizable in "crisis" firms: 46 percent of these had not yet attained a positive profit situation at all, 25 percent were fluctuating unstably between years of profit and loss, with losses predominating in 1993. 29 percent were already in the black, mostly after having incurred high losses relative to their turnover in the initial years. Crises are often characterized not only by negative trends but often also by **inconstant** or **unstable business situations**. Six firms in the crisis group who had made a profit in the first few years were not able to stabilize this good starting position and in subsequent years either fluctuated between profit and loss, or their annual results consistently went into the red, resulting in extremely high negative return-on-turnover and high cumulative losses since foundation. NTBFs that have achieved positive returns for the last two or more years in succession can be expected to have left the critical phase behind them. Also under individual scrutiny, the development of results in the **crisis-free** NTBFs presents a quite different picture: with a few exceptions, these firms had achieved profits or at least a break-even situation by the time of the interview (1993).

Among the 27 crisis enterprises there are also (a few) NTBFs with a positive business development that underwent dangerous crises in their first few years involving relations between partners or management, without business figures being affected. Since some NTBFs also experienced crises despite good business development, it would therefore appear that turnover and profit are not too appropriate up to a point as criteria for the clear identification of crisis situations.

Analysis of the business development of the 42 NTBFs, based on data on turnover and profit development, confirms that non-continuous development of turnover and high or continually-increasing losses are all clear indications of a crisis. On the other hand, it cannot be concluded conversely that an enterprise with continuous positive development has not had to cope with any crises. However, crises do occur less often. This underlines the fact that the analysis of success and crisis is complex and cannot be adequately expressed in terms of simple, straightforward quantitative indicator such as the ones used here.

3.4   Qualitative Crisis Symptoms

The following section describes the non-quantifiable symptoms of crisis and their constellations, relating them to the 27 crisis enterprises from the sample as identified above. A distinction is made between endogenous and exogenous crisis determinants.

Endogenous crisis determinants

In the majority of the 27 crisis enterprises, the NTBF's **preparations for foundation and its strategic orientation** can be considered inadequate and not really appropriate for its aims. This is particularly true of market aspects, i.e. the investigation of customer requirements and the competitive situation.

Contrary to expectations, the incidence of crises among **team foundations** in the 42 NTBFs was not found to be lower than in "single" foundations. Just as in the single foundations, numerous and substantial deficits in **management and leadership skills** were found. Generally speaking, no noteworthy activities aiming to compensate these deficits were in evidence before foundation. In seven "crisis" firms with several active founders, founder-partners left the firm following quarrels and conflicts about the "right" business or problem-solving strategies. These disagreements were frequently triggered by problems resulting from the fact that founders were not able to deal adequately with commercial and marketing tasks, but that the other partners did not take action to combat this inadequacy early on. This generated crises that threatened the survival of the firm. In **"single" foundations**, due to the low degree of delegation of work, all management tasks - and therefore all deficits and problems - usually tended to be concentrated in one person. Moreover, there was no discussion partner to provide a corrective element.

Also due to their size, the 42 NTBFs tend to be characterized by a weak **organizational structure**. Particularly in the crisis firms, the absence of a functioning **controlling and accounting system** often led to an inefficient deployment of limited resources, and a failure to recognize threatening crisis situations until too late.

The **financing crises** that frequently occur in NTBFs mainly have their causes in **erroneous assessments** of the time and capital required, the qualification requirements at market introduction of the innovative products, and in over-optimistic expectations regarding market response. Thus financing bottlenecks occurred as a consequence of deficits in other areas. These often forced founders into a vicious circle of dwindling company capital and loss of creditworthiness. Only if investors or credit institutions assessed future business development positively did additional capital flow into the NTBF. In the other cases, investment was postponed, planned activities were reduced or the firm was fundamentally re-structured.

Founders of firms that suffered from crises later on had mostly underestimated the influence of systematic **marketing** on the success of market entry, not giving sufficient attention to a detailed marketing strategy or to a sales orientation of the whole firm. In addition, there were large deficits in marketing and sales know-how. In almost half of these NTBFs, the firm's offer consisted of only one product line; the capital necessary for further product diversification was usually lacking.

Exogenous crisis determinants

The main factor to be emphasized here is the difficult state of the economy, affecting the crisis firms during their build-up phase. A substantial proportion of them were seriously disadvantaged as a result of the economic trend in Germany, since market entry was rendered much more difficult by the cautious buying behaviour of potential customers, meaning that turnover and profits did not develop according to plan. For some crisis firms, market entry was impeded in addition by state or non-state authorisation procedures, norms, rules and standards, whose importance and effects were underestimated by the founders. Moreover, some NTBFs were con-fronted by unforeseeable events outside their sphere of influence (e.g. the death of a founder, untrustworthy behaviour of an investor, customers failing to pay), which either caused crises or aggravated them.

3.5   Types of Crises

The empirical results on the courses of crises in NTBFs suggest a crisis typology linked with the individual phases of development in the process of building up these firms. Four types of crisis (start-up/build-up crises, market entry crises, survival crises and growth crises) can be identified. Each type can have numerous different characteristics which, in real instances, do not all appear together. The main causes of crises in all phases of development are: information and know-how deficits in management, combined with the highly demanding tasks associated with building up a completely new business unit, and the insufficient availability - or insufficient involvement - of external resources. Because of the composition of the sample be-ing considered, the crises in the NTBFs interviewed were predominantly in the start-up/build-up phase, up to and including crises at market introduction. Due to the young age of the NTBFs in the sample, crises triggered by particularly rapid growth were not observed.

3.6   Behaviour of Capital Investors and Capital Recipients in Crisis Situations

Relative frequency of crisis firms and types of investors

Among other criteria, capital investors can be classified according to whether or not they offer NTBFs management support in addition to their capital (on the typology of investment companies, cf. Harnischfeger/Kulicke/Wupperfeld 1993). Seed and venture capital companies emphasize that they have servicing and consulting concepts that are suited to the needs of technology-based firms in the early phases of their development (on this aspect, cf. Wupperfeld 1994b; Wupperfeld 1996). Thus they claim to offer NTBFs special "value added", and consequently better chances of success than other investors, whose portfolio support concepts are not tailored to the needs of start-up enterprises, or who confine themselves mainly to making capital available (for instance, the business investment companies of banks, insurance companies and enterprises and, in Germany, the "Mittelständische Beteiligungsgesellschaften", or MBGs, which are investment societies founded with the purpose of supporting small and medium industry).

Whether or not an explicit agreement has been made about consulting and management support, it must be in the basic interests of all providers of equity capital to support the firms they invest in at times of crisis. This applies particularly to silent partners. Fees on capital provision leave very little scope for counterbalancing the full loss of an engagement. Investors who contribute directly to the company capital and take their returns from the increase in value of the shares can counter a setback much more effectively with an appropriate portfolio support policy.

Although the database from the interviews with 42 NTBFs gives a detailed picture of the interrelations between types of investors and the incidence of crises in firms (see Figure 6), it is too small-scale to be used as confirmation of the theory that the occurrence and outcome of crises in an enterprise are dependent on the type of the investing partner. If, for this sample, capital investors are grouped according to whether or not an explicit agreement has been made with them regarding management support, the distribution of "crisis" firms and crisis-free firms is fully homogeneous - in other words, no correlation is observed between absence or presence of management support and the occurrence of crises.

Figure 6:   **Incidence of crises and type of capital investor (lead investors only)**

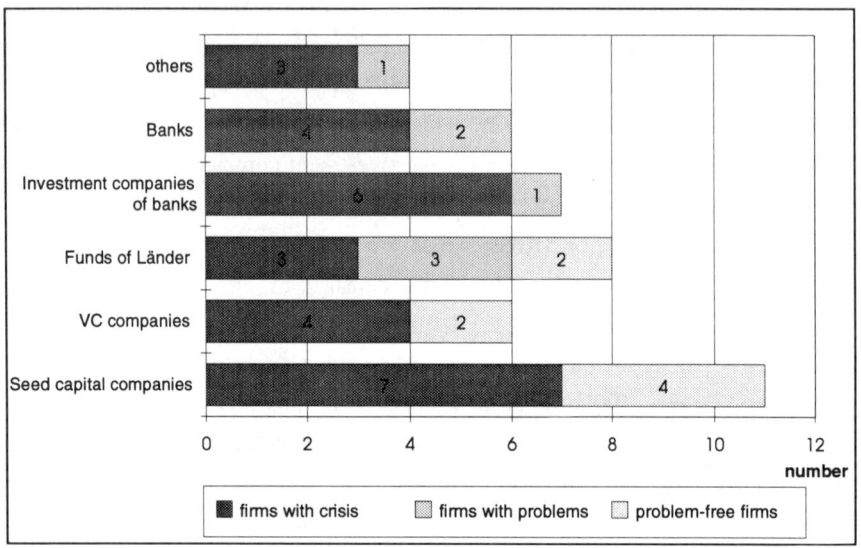

Crisis enterprises and investment volume

Figure 7 shows the comparative distribution of investment volume for NTBFs experiencing crises and crisis-free NTBFs (the relevant information was available for 24 of the 27 crisis firms and 13 of the 15 crisis-free NTBFs). In the latter, lower amounts of capital had flowed into the firms than in the others. Nevertheless, every fourth "crisis" firm had at its disposal risk-bearing capital amounting to three million DM or more. This finding can be interpreted as indicating that substantial amounts of (investment) capital had been brought in to overcome crises.

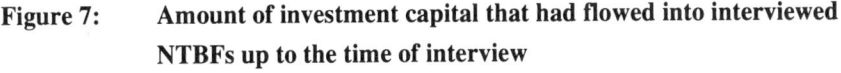

Figure 7: **Amount of investment capital that had flowed into interviewed NTBFs up to the time of interview**

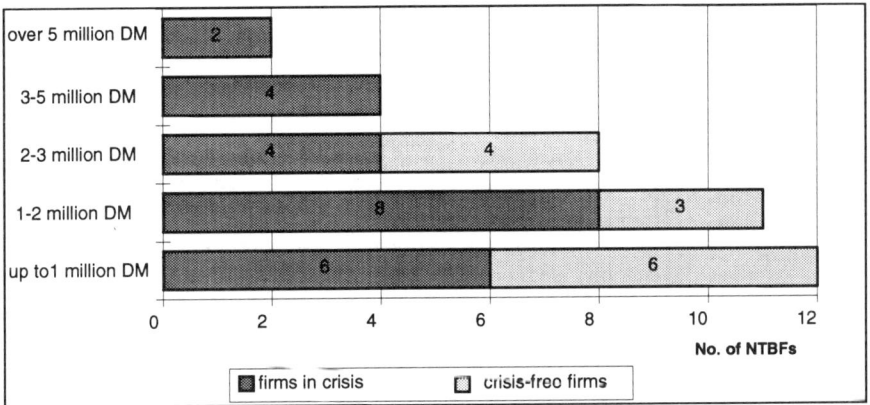

The role of capital investors in combating crises from the viewpoint of the NTBFs

This section follows up the question of how capital investors behaved in the event of a crisis, and particulary how they performed crisis management in their portfolio enterprises. There were substantial hindrances to a valid analysis of this aspect, due to the possibilities for obtaining information: interviews with founders directly affected by crises can only reflect a subjective view of the causes and development of crisis situations.

The possible roles of capital investors in actively supporting NTBFs are presented in Table 16. The roles are performed with varying intensity depending on the type of investor, since suppliers of capital pursue individual investment and support strategies. Venture capital and seed capital companies explicitly emphasize their intensive support of NTBFs in all questions concerning the building-up of the firm. By contrast, investors who are silent partners (e.g. societies for the promotion of small and medium industry such as the German MBGs) or provide conditional loans (banks), tend to see their role as "hands-off", and observe the development of the firm from a distance - except in crisis situations, or under exceptional business circumstances. These divergent servicing policies have to be taken into account when assessing the role played by the investment company.

Table 16:   Roles of investment companies in the servicing of NTBFs

Role	Activity	Time dimension
Controller	Monitors business development and finance situation	permanent task
Mediator of contacts	Mediates contacts to other investors, cooperation partners, consultants, potential customers, suppliers, lawyers, tax advisers and other know-how sources	occasionally
Mediator of information and know-how	Mediates knowledge of business methods and information - e.g. about the market, competitors, customers - through personal contacts, seminars, discussion rounds between the portfolio firms; one aim is to enhance the skills and qualifications of the founder (help towards self-help).	occasionally
Sparring partner	Competent discussion partner, e.g. in the critical analysis of strategies or operative problems	occasionally
Business consultant	Carries out extensive consulting tasks, such as e.g. re-organization, introduction of a new EDP system, market studies	in specific situations and occasionally
Temporary manager	Takes on entrepreneurial tasks, e.g. in crises	in exceptional sitautions

Source: Wupperfeld 1994a: 132

From the viewpoint of the firms interviewed, a very heterogeous picture emerges with regard to the support services actually performed. Three groups of NTBFs can be distinguished: the first group mentions the excellent crisis management performed by their investors, whereas the second emphasizes negative effects of the management, making the crisis worse. The group in the middle did not receive any support to speak of. Whereas the expectations of crisis-free NTBFs regarding their capital investors were largely fulfilled, firms with crises fairly often experienced serious discrepancies between their expectations of support and the support they received. The same is true of the evaluations of the usefulness of consulting. Particularly in the areas of marketing and sales, where the biggest problems occurred, there was a striking lack of congruence between the demand and the support received from the investor.

4.   Summary and Conclusions

The survey of 42 NTBFs receiving subsidized investment capital under the BJTU pilot scheme shows that the probability of a serious crisis - understood as a constellation of disruptive parameters that may endanger the existence of the firm - occurring in the first few business years of a new technology-based firm is very high. The figure of 27 NTBFs out of the group of 42 means that a relatively large proportion of firms had had to face crises in the course of their existence so far. On the other hand, in eight NTBFs the firm was built up without any substantial problems. Although the remaining seven NTBFs experienced difficulties, these did not amount to crises. Analysis of the causes and paths of crises underlines that multiple criteria are needed to describe them. Crisis situations cannot be adequately recognized by exclusively regarding the development of turnover and profit. It would seem that detailed analysis of the firm's development so far is essential in order to make reliable statements about crises that may arise. This is particularly true for the early diagnosis and anticipation of crises.

Serious problems in one functional area of a firm (e.g. in sales) usually precipitate a chain-reaction crisis affecting all areas of the firm's activity. Only a few NTBFs successfully overcame crises in the period of observation, which coincided with the worst recession phase of the German economy for the past few decades. However, among the 27 firms that experienced crises there are some examples of successful "turn-arounds". **Moreover, a crisis cannot be generally interpreted as a failure of the firm; rather, it is a typical manifestation of the build-up of that complex business unit, a new technology-based firm.**

In almost two-thirds of all cases the crisis occurred during the start-up and build-up phases or at market introduction of the innovative offer. This underlines the high risks to which a technology-based firm is exposed. The second most frequent types of crisis to occur were serious problems in establishing a permanent position on the market and in making adequate profits.

The interrelations between causes of crises and their impacts is higly complex. In reality, any of the risk factors affecting NTBFs may precipitate a crisis. There is a complicated hierarchy of causes and effects; generally, several causes combine. Thus attempts at monocausal explanations are self-defeating. In the method of analysis selected, it proved practically impossible to isolate and weight the individual factors that were under consideration as crisis-precipitating factors.

Nevertheless, certain factors can be **excluded** as primary causes of crises for the NTBFs supported through the BJTU scheme, from the point when they took on subsidized investment capital: for instance, financial bottlenecks, since due to the financing by equity and the other sources of financing that this opened up, the NTBFs seemed to be adequately equipped with capital at least for the next two to three years. If a financing crises arose in spite of this, it had to be the consequence of some other deficit such as e.g. wrongly assessing the real need for capital and consequently raising too little investment capital.

A financing crisis often leads to a vicious circle of dwindling company capital and loss of creditworthiness. An NTBF can only break this vicious circle on its own if there is a rapid positive development of profits, so that its financial standing with investors also improves. However, this situation was an exception among the NTBFs interviewed, even if a very good product was being offered on the market: NTBFs did not generally reach the profit threshold before the fourth year of business and their profits were not yet sufficient to compensate the consumption of company capital. Thus even successful NTBFs needed additional capital to finance market diffusion (i.e. to build up sales and manufacturing). According to the experience of the NTBFs interviewed, the possibility of receiving substantial amounts of risk-bearing **capital for the financing of the marketing, expansion and growth phases** of new technology-based firms should constitute an important component of both promotion policy support and of the financing support given to these firms.

Technical problems in the realization of the innovation project hardly ever appeared as the causes of crises. Founders are well-qualified in this field. However, the area of R&D was relevant for the occurrence of crises, in the sense that the time and expense of development were often seriously underestimated, a mistake which constituted an important cause of financing bottlenecks later on. This demonstrates the need for investors to **introduce more milestones and progress controls**, in order to combat the tendency for capital to get used up at an early stage when the limits of expenditure set for the planned development are exceeded - a tendency which has very negative strategic impacts.

Crises often have their causes in the type of preparations for the foundation. According to founders' own statements, in more than half of the firms experiencing crises, foundation had only been systematically prepared up to a point. However, in-depth interviews led to the assessment that, very often, insufficient care and thoroughness had gone into elaborating the business concepts. The figures on which they were based later proved not to be sufficiently reliable, so that these business concepts could not function as approximate guidelines or bases for planning, against which the success of the firm could be measured. In many cases investors and banks, which generally required a business plan to be elaborated and cooperated in its elaboration, obviously did not recognize or correct these deficits. This leads to the conclusion that investors and experts examining the business concepts should examine very carefully the quality and details of the preparations for foundation, paying particular attention to what sources the figures in the business plan came from, and how they have been checked.

All the factors identified as relevant to crises appeared in different constellations in individual cases. They were grouped into several crisis fields, or areas of influence, and illustrated by examples. In all these crisis fields, qualification and management deficits in the firm's management can be identified as the main cause. However, it should be borne in mind that the skills required to build up a completely new business unit in dynamically changing markets, against the background of a serious economic recession, are so complex that deficits on the part of founders are practically inevitable. Promotion of the founding of NTBFs aims primarily to encourage towards the step of self-employment people who are, first and foremost, excellent technicians and developers with an innovative product idea, and possess at least a basic knowledge of markets or commerce. It is difficult to assess beforehand the extent to which they also have the potential to mature into successful entrepreneurs in the course of building up their business. **In view of the complex influences at work, crises have to be regarded as typical manifestations in the process of building up new technology-based firms.** In the BJTU pilot scheme, existing deficits in experience and information in non-technical areas should be compensated by the provision of support services by investment companies. The only statement that can be made for the NTBFs investigated here is that only in some cases were support services rendered which contributed to overcoming crises.

When making an overall evaluation of the causes and outcome of crises in NTBFs on the one hand, and of the support services provided by investors on the other, the following main conclusions can be drawn: since deficits in the know-how and experience of the founders are important precipitators of crises, since investment capital is often raised only to overcome a crisis that has arisen, and since, even at this stage, investors only provide effective management support up to a point, promotion targeting NTBFs should be extended to include qualification measures for potential founders, **before** they take the step of going self-employed. Unlike the the situation in the USA, where relevant "entrepreneurship courses" are established at many universities, in Germany there is still a lack of training and further training courses tailored to the target group of "technology-based firm foundations" and imparting, especially in the area of marketing and sales and the elaboration of business strategies, detailed know-how that goes beyond the stage of basic knowledge. Offers of this kind should follow a twofold plan: on the one hand, as a medium- and long-term strategy, they should be included as part of the training of engineers and scientists at universities and institutions of higher education. On the other, they should be available as further training to people with professional experience working in research institutions and enterprises, who can be directly considered as potential founders. This could lead to the foundation of greater numbers of technology-based firms, thus increasing the technology transfer from existing institutions and leading to a general expansion of the innovative "Mittelstand" (small and medium-sized industry).

5. Bibliography

Albach, H. (1988): Maßstäbe für den Unternehmenserfolg, in: Henzler, H.A. (Ed.): Handbuch Strategische Führung. Wiesbaden, 69-83.

Bayer, K. (1990): Beratung und Betreuung junger Technologieunternehmen - Erfahrungen aus dem Modellversuch TOU. Working paper. ISI: Karlsruhe.

Gerybadze, A. assisted by Kulicke, M. (1990): Erfolgsbedingungen und -kriterien für junge Unternehmen, insbesondere für technologieorientierte Unternehmensgründungen. Working paper. ISI: Karlsruhe.

Gutenberg, E. (1989): Rückblicke. Unpublished manuscript. Köln 1983, subsequently reproduced in: Albach, H. (1989): Zur Theorie der Unternehmung. Schriften und Reden von Erich Gutenberg aus dem Nachlaß. Berlin, Heidelberg, New York, 1-109.

Harnischfeger, M./Kulicke, M./Wupperfeld, U. (1992): Zum Stand des Modellversuchs "Beteiligungskapital für junge Technologieunternehmen" (BJTU) - Zwischenbericht zum 31.12.1991. Working paper. ISI: Karlsruhe.

Hemer, J./Kulicke, M. (1995): Krisen in jungen Technologieunternehmen - Eine empirische Analyse der Krisenverläufe von im Modellversuch "Beteiligungskapital für junge Technologieunternehmen" (BJTU) begünstigten Unternehmen. Working paper. ISI: Karlsruhe.

Klein-Blenkers, F. (1976): Insolvenzursachen mittelständischer Betriebe. Eine empirische Analyse. Göttingen.

Kraft, A. (1986): Potentialanalyse des Unternehmens in der Krise. In: Schimke, E./Töpfer, A. (Eds.): Krisenmanagement und Sanierungsstrategien. Landsberg a. Lech, 16-27.

Krüger, W. (1988): Die Erklärung von Unternehmenserfolg: Theoretischer Ansatz und empirische Ergebnisse. In: Die Betriebswirtschaft (48), 27-43.

Kulicke, M. (1990): Modellversuch "Förderung technologieorientierter Unternehmensgründungen" (TOU). Zwischenbilanz zum 31.12.1989. Working paper. ISI: Karlsruhe.

Kulicke, M. (1994): Ausfallraten junger Technologieunternehmen im Modellversuch "Förderung technologieorientierter Unternehmensgründungen" (TOU). Auswertungen zum Stand 30.6.1994. Working paper. ISI: Karlsruhe.

Kulicke, M. assisted by Bayer, K./Walter G.H. (1991): Modellversuch "Förderung technologieorientierter Unternehmensgründungen" (TOU). Zwischenbilanz zum 31.12.1990. Working paper. ISI: Karlsruhe.

Kulicke, M. et al. (1993): Chancen und Risiken junger Technologieunternehmen - Ergebnisse des Modellversuchs Förderung technologieorientierter Unternehmensgründungen (TOU). Heidelberg.

Kulicke, M./Hemer, J./Wupperfeld, U. assisted by Traxel, H. (1994): Zum Stand des Modellversuchs "Beteiligungskapital für junge Technologieunternehmen" (BJTU) - Zwischenbericht zum 31.12.1993. Working paper. ISI: Karlsruhe.

Marré, G. (1986): Controlling in der Krise. In: Schimke, E./Töpfer, A. (Hrsg.): Krisenmanagement und Sanierungsstrategien. Landsberg a. Lech, 62-76.

Paul, H. (1985): Unternehmensentwicklung als betriebswirtschaftliches Problem. Ein Beitrag zur Systematisierung von Erklärungsversuchen der Unternehmensentwicklung. Frankfurt am Main, Bern, New York.

Schimke, E./Töpfer, A. (Eds.) (1986): Krisenmanagement und Sanierungsstrategien. Landsberg a. Lech.

Töpfer, A. (1990): Insolvenzursachen, Turn-around, Erfolgsfaktoren. über existenzbedrohende Stolpersteine zum Unternehmenserfolg. 1. Teil. In: Zeitschrift für Organization, Nr. 5, 323-329.

von Wichert-Nick, D. assisted by Kulicke, M. (1994): Ökonomische Entwicklung und Unternehmensstrategien junger Technologieunternehmen. Ergebnisse einer Befragung von im Modellversuch Beteiligungskapital für junge Technologieunternehmen (BJTU) begünstigten Unternehmen. Working paper. ISI: Karlsruhe.

Wupperfeld, U. (1993): Mißerfolgsfaktoren junger Technologieunternehmen. Working paper. ISI: Karlsruhe.

Wupperfeld, U. (1994a): Strategien und Management von Beteiligungsgesellschaften im deutschen Seed-Capital-Markt - Ergebnisse einer empirischen Untersuchung von 33 Beteiligungsgesellschaften und Banken. Working paper. ISI: Karlsruhe.

Wupperfeld, U. (1994b): Die Betreuung junger Technologieunternehmen durch ihre Beteiligungskapitalgeber. Empirische Untersuchung. Working paper. ISI: Karlsruhe.

Wupperfeld, U. (1996): Management und Rahmenbedingungen von Beteiligungsgesellschaften auf dem deutschen Seed-Capital-Markt. Frankfurt am Main.

Consulting for New Technology-Based Firms

Marianne Kulicke

1. The Importance of Consulting in the German Federal Pilot Schemes for New Technology-Based Firms

Since the 1980s, the German Federal government's promotion measures for new technology-based firms (NTBFs) have not been concerned with purely financial support, but have also explicitly contained an advisory component (cf. Kulicke et al. 1993). In the pilot schemes "Promotion of New Technology-Based Firms" (TOU) in the old and the new federal states, the task of giving advice and support to firms benefiting under the scheme was entrusted to special technology advisory offices offering both technical and commercial know-how, with the intention of providing continuous support for the newly-founded firms during the promotion period.

Particularly with regard to this consulting component, the Federal Ministry of Education, Science, Research and Technology (BMBF) was breaking new ground with its first pilot scheme in the old federal states (TOU/ABL). Within a framework of tasks which was contractually fixed, the technology advisory offices retained freedom to design their own consulting and support services. The support provided covered the whole start-up process of a NTBF, from preparations of the foundation, through development of the product or process up to market introduction.

In addition, under Promotion Phase I of this pilot scheme, consulting services from the private sector were also brought in from time to time. These related to the elaboration of the basic business plan (cf. Mayer et al. 1986). In this way, a structural policy component was built into the support for NTBFs. Thus the TOU/ABL pilot scheme had the aim of "sensitising" the suppliers of innovation services to the specific needs of NTBFs. This promotion under Phase I created a much greater demand than before for foundation consulting in innovative projects. Its purpose was to enable qualified consulting institutions to develop consulting offers which were specially tailored to the requirements of this customer group. If successful, this very ambitious structural policy would have meant that a market segment (albeit a small one) specifically offering consulting services to NTBFs became permanently established in the German consulting market.

Particularly the involvement of private consultants, but also the consulting by special technology advisory centres, did not produce the intended effects (cf. Bayer 1990). With regard to the former aspect, it became apparent that the target group of NTBFs was much too small, and its consulting requirements much too heterogeneous, for these firms to represent a lucrative business field for private consultants. Above all, not every need for consulting resulted in a corresponding consulting demand. Moreover, the costs of qualified consulting far exceeded the financial possibilities of newly-founded firms. There were several reasons why consulting by the technology advisory offices did not lead to the hoped-for success. Firstly, the heterogeneity of founders' requirements also played an important role here, particularly their need for support in sales and marketing questions. However, one important snag in the activity of the technology advisory offices turned out to be that they were limited to providing support in a purely advisory capacity, so that if a firm was developing in a seriously wrong direction, the most they could do to counter this development was to block the payment of subsidies. Moreover any advisor, whether privately or publicly financed, is only affected to a limited extent himself by the consequences of his advice.

The follow-on measure of the TOU/ABL scheme, the pilot scheme "Business Investment Capital for New Technology-Based Firms" (BJTU), adopted a completely different way of financially supporting the target group, and concomitantly aimed at another form of management support (see BMFT 1989; Bräunling et al. 1989). Instead of public subsidies and guarantees for longterm loans, and subsidized technology advisory offices, the follow-on scheme supported the provision of capital by business investment companies, credit institutions, enterprises and private investors; the intention was that these investors would then support the recipient firms in all matters relating to building up the firm, either by advising them themselves or by bringing in a third party. This approach was based on the consideration that investors have a lasting self-interest in the successful development of their investees and will accordingly support them in non-financial aspects as well (cf. Schröder 1992). These support services themselves were not subsidized by the BMBF. Rather, the investors were free to finance them with their income from investments, or possibly to derive income from consulting agreements with the recipients of capital.

The following chapter first presents the specific needs of NTBFs for advice and support. Since it was primarily in the TOU/ABL pilot scheme that analysis centred on this issue, the presentation is based on the experiences gathered in this promotion measure (cf. Kulicke et al. 1993). Information gathered on the consulting services offered by investment companies and NTBFs assessments of them relates to the BJTU pilot scheme (cf. Wupperfeld/Hemer 1995; Wupperfeld 1996).

2. The Consulting Needs of New Technology-Based Firms

The most important item of knowledge gained from the consulting component of the TOU/ABL pilot scheme was that the **extent and intensity of the need for consulting of NTBFs,** and thus also the spectrum of their requirements for consultants, are **very heterogeneous**. There can be no question of a uniform need for consulting for all types of NTBFs (cf. also Baaken 1989).

These firms' need for support derives from the specific requirements of the start-up process (Kulicke 1990) and from the individual strengths, weaknesses and personality structure of founders. The content of consulting, and its intensity over time, are a function of the relation between the capabilities and resources that are required, and those that are available. In favourable situations, the capabilities of the founder and the requirements coincide, and there is no need for consulting. In unfavourable situations, the disparities are so great that an external person or institution has to be brought in with the task of taking on, at least temporarily, the role of co-entrepreneur.

However, the need for non-material resources does not remain constant from the time of formal founding to market establishment (Kulicke/Gerybadze 1990; Picot et al. 1989). It is largely determined by the aims of the founder in founding his firm. The need and type of external support services will also depend on the phase of development of the enterprise, on the course of start-up and development in the NTBF, and on the technical and market risks associated with the product or process to be developed.

The extent to which the consulting requirements of new and established (technology-based) firms differ, if at all, depends on the content of the consulting and the level at which it takes place. If strategies and fixing the characteristics of a new business unit are the central issues, the lack of track record data and problems in assessing technical and market risks require a fundamentally different approach. The more strongly consulting is concerned with operative aspects (e.g. introducing an information and control system, optimizing manufacturing equipment, seeking managerial staff via a personnel consultant, etc.), the smaller are the differences between new and already-existing enterprises.

In NTBFs, the **personality of the founder** plays a central role (Unterkofler 1989; Gerybadze/Kulicke 1990). Consultants must be prepared, and able, to adapt to the founder's mode of thought. They should be capable of assuming different roles. The primary role is that of the "sparring partner" or discussion partner. From their previous professional activity, quite a lot of the founders of NTBFs supported under the pilot schemes already had experience in the management of big development projects. However, experience of this kind is not necessarily directly transferable to the management of marketing activities, and vice versa - even if some of the basic work routines in the planning and execution of an R&D project and a marketing project are identical. Apart from the non-comparability of their available resources, the decisive difference between the situation in a large enterprise and a new firm foundation is the fact that the former has an established organization for building up and running the enterprise, with clearly-structured competencies and responsibilities. In the newly-founded firm, however, a functioning organizational structure first has to be created and then adapted as efficiently as possible to the changing requirements that arise during the process of growth (Gerybadze 1988).

When founders' know-how is concentrated in the technical field (Kulicke et al. 1993) there is generally a great need for learning, i.e. it is necessary for founders to acquire knowledge and skills they did not previously possess. Frequently, however, this is not a matter of a serious deficit, but just a lack of experience which can be successively acquired during the first few years of business while dealing with the demands involved in building up the firm. Learning is needed in the commercial aspects, particularly marketing, but also the planning of finance, operative controlling and the acquisition of capital. These needs can be met either by "learning by doing", or by bringing in a consultant; in the results of the TOU/ABL pilot scheme, neither way emerged as being more efficient than the other.

One of the **main characteristics of the consulting requirements in NTBFs** is the need for **integrated** consulting. Consulting has to relate to the whole firm and the totality of its resources, and has to anticipate the steps the enterprise will be taking in the future. The TOU/ABL pilot scheme makes it clear that in order to be suitable for NTBFs, consultants not only have to possess know-how in business and sales aspects, but must also already have knowledge of the NTBF's area of technology, or must be in a position to acquire it quickly. In addition, in order to make a valid assessment of market chances and develop market strategies, basic knowledge of the relevant sector, competition structures and customer requirements are necessary. Consulting may be used for the following purposes:

- to **elaborate a business concept** at the beginning of the foundation process,

- in the context of the planned development of the business as a whole, only to **establish strategies for individual functions** (e.g. building up the marketing organization, ways to open up foreign markets, division of tasks in manufacturing),

- from time to time, to **mediate** relevant information, set up contacts to external R&D capacities, investors of venture capital, sales intermediaries, etc., or

- for (management) support in operative areas of the firm.

In extreme cases, a NTBF may require services in all four of these aspects. If, ultimately, the consulting does not generate perceptible learning effects, the question arises as to the founder's suitability as an entrepreneur. Only for a few of the NTBFs supported under the TOU/ABL pilot scheme this unfavourable constellation held true, most of these firms failed.

The need for consulting varies in the individual phases of the firm's development according to the tasks and problems to be mastered (Kulicke 1990). **Before commencing the development project**, the most important elements of consulting are the conception and examination of the business strategy to ensure that it is complete, appropriate and realistic. At this point, for instance, questions relating to legal status, success prospects of the R&D goals, need for capital, market chances and stability of demand, etc., need to be clarified. This situation is characterized by the great number of degrees of freedom in decision-making, combined with a high degree of uncertainty in assessments and strong interactions between the variables.

This is the point of application for consulting which assists the founder in realizing his concepts and ideas, examines their consistency, pinpoints their weaknesses and thus contributes, in an iterative process, to the elaboration of a solid and realistic business plan. However, it is prerequisite for this process that the relevant parameters for decision-making and planning are made known as far as possible to the founder. The consultant takes on the role of a sparring partner, a critical questioner of assumptions. Consulting is based primarily on the existing knowledge and information of founder and consultant. The consultant's most important instruments are broad experience of cooperation with NTBFs, a knowledge of the problems that typically occur in the development of these firms, and the ways in which they are generally solved.

Often, however, a need arises for complementary measures, since the information base is not adequate to arrive at a valid assessment of the success prospects of the business plan under consideration. Activities of this kind include, for example:

- market research on future customer requirements and the behaviour of competitors,

- an expertise of the planned technical solution by an external technical expert,

- the mediation of contacts to financiers and

- assistance in seeking and selecting other "key persons" for the start-up (either as employees or as co-founders).

In this consulting situation, as well as fulfilling the functions of a "sparring partner" and a "source of know-how", the consultant may also have to assume the tasks of information clearing (What further information is needed? Who can provide it?), and mediation of contacts. He may possibly also have to act as a moderator in discussions with various experts, in the process of elaborating an appropriate business plan.

The innovative projects of NTBFs are associated with very high market risks. These firms very often need advice on all aspects of market entry and market processing, both regarding demand requirements and the offer to be developed. Since a consultant cannot be familiar with the markets of all the firms he advises, effective assistance has to be based on the application of certain methodologies, in order to identify the risk factors on the market side in this type of innovative project and, on this basis, work out a promising "line of attack" together with the founder or with a third party.

While **carrying out the development project**, the development line is usually largely fixed. When pursuing this development line, the firm ties down its resources and its capacities. It loses its ability to react flexibly to changes in the premises on which the concept was based. Towards the end of the development work at the latest, but often in parallel with it - particularly if other products or services are being marketed - the various functional areas of the firm are reified, thus making further demands on the management skills and capacities of the entrepreneurs. The first point of application for consulting during this phase may be strategic and operative controlling, i.e. permanent monitoring of the basic premises of the business concept to check that they are still realistic, check the progress of the development and monitor the use of resources. Where deviations are found, decisions have to be made on whether to react, and how, particularly with regard to possible reactions of the firm's business partners. Further points of application for consulting at this stage may be questions relating to operative management (e.g. the search for personnel, personnel management, the development and adaptation of organizational procedures) and the future-oriented initiation and development of relations in the firm's environment. This applies primarily to sales, but may also be relevant for production of the newly-developed products.

In the **product development phase**, the new firm receives little or no feedback information from its markets. Here, support from strategic controlling is often a starting-point for consulting (i.e. monitoring of competitors' behaviour, early inclusion of changes in demand requirements into the product being developed). For founders, it is objectively important at this stage (though not always subjectively perceived as important) not to lose sight of the premises on which their development work is based, or its market orientation.

In the **market entry phase** the need for consulting centres on operative aspects, provided that the firm has been sufficiently market-oriented in its previous phases. For NTBFs supported under the pilot schemes, market entry proved to be critical from a financial viewpoint, if they did not already have turnover from other products or services. Logically speaking, this financial situation calls for fast establishment in the market place. If market entry is delayed for financial reasons, there is a danger that competitors will catch up with the temporal competitive edge and siphon off the innovator's profits. A need for consulting may arise here either with regard to strategic questions (e.g. opening up the market, timing of marketing activities, raising sufficient capital) or operative questions (e.g. form of distribution agreements, presentation at fairs, pre-financing of orders, search for - and cooperation with - pilot customers, etc.).

The success of consulting in the start-up phase of a NTBF should also be manifest in the ability of the new business unit's management, after passing through these three stages, to identify and solve conceptual and strategic problems itself. The quality of the advice given is, of course, only one component of the successful outcome of consulting. The other component is the adaptation and learning capacity of founders in the critical first few business years of their enterprise. Only when these development phases have been successfully traversed do the consulting needs and requirements of NTBFs begin to come closer to the needs of innovating small and medium-sized enterprises in general.

During product development and market entry, strategic errors from earlier phases often become apparent. This results in a need for intensive crisis management support, comprising strategic and operative advisory elements combined with direct assistance in negotiating with the house bank.

This description of a frequently-occurring need for consulting in NTBFs, from the refining of the business concept through to market establishment, gives indications of general **fields of activity for consultants advising founders**. NTBFs show different patterns of development, which are expressed in highly divergent cooperation patterns with exogenous partners, including consultants (von Wichert-Nick/Kulicke 1994).

The TOU/ABL pilot scheme also shows that the experiences of private consultants in dealing with established firms can be transferred relatively easily to the operative management problems of NTBFs. The strategic level, however, includes elements which are specific to the group of NTBFs. It is rarely possible to gather analogous experiences of these specific elements in a different context. It is beneficial for consulting at a strategic level if the consultant has a technological focus. By contrast, crisis management is facilitated if the advisor is embedded into the network of actors and institutions in the region where the NTBF is located.

Due to the limited potential demand arising each year from newly-founded technology-based firms, it is understandable that private and public consulting and advisory institutions do not specialize in this segment or specifically build up their knowledge and experience potential in this direction. It is not to be expected that demand will expand significantly in the near future, as firms of this kind often do not have sufficient means in the first few years following foundation, or are not prepared to incur the costs of consulting tailored to their specific problems.

From the experience gained in the TOU/ABL pilot scheme, the following **elements of an appropriate support concept meeting the needs of NTBFs** can be derived:

1. Basic consultation of the enterprise over a long time horizon, to provide assistance in ensuring a sound and realistic strategic basis for the firm and help with strategic controlling in the course of development.

2. An advisory relationship, also long-term, addressing questions of operative management and characterized more by the breadth than the depth of the demands made on the advisor's know-how.

3. Consulting on clearly-defined individual tasks, particularly in the area of marketing. These tasks tend to occur after the stage of concept definition. The advisor should be a proven specialist using "state of the art" methods in the relevant field. A lasting association with the enterprise is not necessary here.

4. To complement these aspects, assistance in excess of usual advisory services *per se* needs to be given in occasional situations which it is critical for the success of the firm to overcome (crisis management; on this aspect, cf. also the chapter in this reader on crises in NTBFs). This support element also seems to be best provided in the context of a long-term relationship between founder and advisor.

The numerous and varied requirements of NTBFs mentioned above make high demands on an "ideal" consulting partner. Since either private or public advisory institutions can only meet these demands in part, the BJTU pilot scheme adopted a different approach in which newly-founded technology-based firms were supported not only financially, but also by a flow of external know-how into the firm.

3. Management Support Provided by Investors for New Technology-Based Firms

3.1 Preliminary Comment

In view of the many different types of business investment companies and credit institutions financing NTBFs under the BJTU pilot scheme, there is a correspondingly broad variation in intensity of support and the areas in which it is given (cf. Wupperfeld 1996). This section describes the fields in which firms participating in the BJTU pilot scheme experienced support, how intensive the support was, and how the firms evaluated it.

For a business investment or venture capital company, management support for its portfolio enterprises only make sense if the usefulness of the support clearly exceeds its cost. Moreover, any fund has only limited management resources at its disposal (both quantitatively and qualitatively speaking). For this reason, the intensity of management support provided by investors for the recipients of their capital varies in its intensity and in its emphasis. **From the viewpoint of the investor**, it may under some circumstances make economic sense to service all or some individual portfolio enterprises less than would be required for their optimal business development. In extreme cases, this may even mean that the management support will give up firms with poor chances, even if there is a danger of their investment being written off completely. **From the viewpoint of the founders of NTBFs**, who are naturally interested in their firms continuing to exist, this behaviour certainly has to be judged quite differently. This leads to conflicts between the aims of investor and investee. However, conflicts may also occur if founders experience over-intensive support as "meddling". Thus the "optimal" extent of support, as perceived by the investment company or by the recipient, may definitely be quite different.

Before going on to describe the empirical results on support services provided by investors, the following considerations should be borne in mind:

- The results are based on subjective statements by founders, who sometimes tend to ascribe successes to their own efforts and failures to their investors, with the consequence that they underestimate the scope and usefulness of the support services.

- In the cooperation between investor and recipient, personal relations play an important part. In crises or differences of opinion, this relationship may suffer serious and lasting damage, with the consequence that founders make negative statements about investors.

- Before concluding the contract of participating some founders did not have a clear idea of the function or aims of a business investment company. They regarded the investor as a sort of supportive institution which was not profit-oriented. This gave rise to unrealistic expectations on the part of these founders.

- The statements made by some founders about support services were remarkably lacking in differentiation. For instance, if an investor's support was inadequate on one occasion, but support had previously been mainly good, it was concluded from this one instance of inadequacy that the support services were generally unskilled.

Since, however, these restrictions only apply in some individual cases, the survey of the BJTU pilot scheme affords in-depth insights into the servicing of NTBFs, and makes possible an evaluation of the usefulness and deficits of the support services. Moreover, it is this subjective viewpoint which is, in point of fact, decisive for the acceptance of business investment or venture capital as a financing instrument.

3.2 The Need for Support and its Main Areas of Application in the Start-up Phase

The need for support in various areas, already apparent from the TOU/ABL pilot scheme, also emerged from the interviews with 42 firms receiving subsidized capital investments under the BJTU pilot scheme (cf. Figure 8, left side). The areas particularly emphasized by founders were marketing and sales. Here, just under 80 percent of founders perceived themselves as confronted by high demands, with deficits at market introduction being regarded as particularly serious. These problems were caused not only by economic trends and by the general barriers to market entry that face NTBFs and technological innovations; they were also due to the founders' lack of experience and lack of contacts. Thus many interviewees emphasized a need for support, not only in developing market strategies and sales concepts, but above all in concrete, pragmatic assistance in operative questions. These included, for instance, mediation of contacts to customers, participation in negotiations with customers, support in preparing for fairs and selecting an advertising agency, etc.

Other important areas in which support is required are financing (e.g. the acquisition of additional investors) and finance management, as well as commercial and management questions. A number of NTBFs explicitly emphasized that here, too, they were in need of practical tips from experienced professionals, continuous corrective monitoring and access to a "business community". Some put forward the view that ideally, support should not commence when problems or crises arose, but should be given beforehand, in order to avoid situations of this kind and improve the firm's chances of success. Only a few interviewees said that they were not in need of support.

This need for support on the part of NTBFs can be compared and contrasted with the **clear points of emphasis in the support provided by investors** (cf. Figure 8, right side).

The most extensive support is given in **financing**: three quarters of the firms received support in this area. Services included help in acquiring loan and equity capital, taking part in negotiations with other investors and aspects of general finance management. Some investors were active in this respect who do not otherwise combine their provision of capital with a detailed offer of supportive services (particularly the credit institutions and the "Mittelständische Beteiligungsgesellschaften" (MBGs), investment companies in Germany founded with the purpose of supporting small and medium industry).

Figure 8: **Need for support compared with support received**

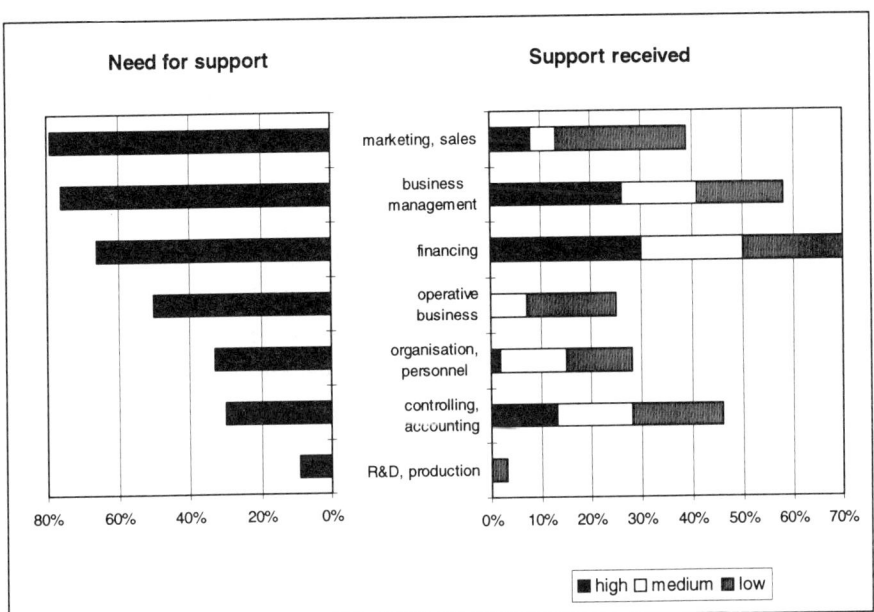

Support tended to centre around financing issues, because business investment capital has primarily a financing function and investment managers generally have high qualifications in this area (Schröder 1992) and are embedded in a well-developed financing network. Moreover, this type of support does not require continuous commitment, but only occasional assistance if the need for support becomes acute (Gorman/Sahlman 1986). The scope of the support only increases when further injections of capital become necessary in order to combat financing bottlenecks or to finance additional investments by the NTBF. Business investment companies prefer sporadic activities of this kind to continuous supportive services, since they are less time-consuming and give them an opportunity to put their expert knowledge to good use (cf. also MacMillan et al. 1988, on the services provided by venture capital companies in the USA). The most intensive servicing was provided by seed and venture capital companies, by private investors and by private firms, whereas business investment companies of the financing sector, credit institutions and the MBGs were less active.

The second focal point of support was the area of **business strategy**. Just under 60 percent of the 42 NTBFs stated that they had received advice concerning strategic business management. Here, the main emphasis was on the development of business strategies and the adaptation of the business plan to market circumstances. This result was also to be expected, since the support philosophy of most venture capital companies consists in assisting their portfolio enterprises primarily in fundamental and strategic matters, while largely avoiding operative aspects of day-to-day business (Cohen 1988; Sapienza/Timmons 1989).

A comparison between the need for support and the emphasis of the support given leads, however, to the suspicion that quite a few recipients of capital did not receive adequate advice in strategic matters. This applies particularly to NTBFs whose main investors were MBGs, but also to firms receiving investments from credit institutions in the finance sector. It should be borne in mind, however, that MBGs usually explicitly offer only limited support. As expected, venture and seed capital companies, on the other hand, were much more active in this area.

Much less support was received by NTBFs in **controlling and accounting.** However, only 30 percent of the NTBFs expressed a corresponding need for support in this area. In cases where assistance was given, investors often began, after conclusion of the investment agreement, to force the build-up of an information and controlling system, which they then made use of to monitor their investment stake.

Statements by founders confirm that NTBFs have an especially great need for support in **marketing and selling.** Nevertheless, over 60 percent of interviewees had received no help whatsoever in this particularly critical area. Only eight percent mentioned intensive support, the rest described it as limited. Thus there is a great discrepancy between the existing demand for advice on marketing and selling and the extent to which it is met. This is also true of some investors whose servicing of portfolio enterprises is intensive in other respects. Recipients of investments frequently stated that venture managers lacked the qualifications necessary to give support in this area (e.g. knowledge of the sector, experience of sales, knowledge of the market).

This deficit may also be attributable to the high costs incurred by concrete, pragmatic assistance (e.g. the search for customers, negotiations with customers and selling partners). Thus support was concerned mainly with conceptual aspects (e.g. the elaboration of marketing or sale plans) and less with putting these concepts into practice. Business investment companies, on the other hand, usually emphasize that they regard operative measures generally, and thus also marketing activities, as fundamentally the task of the entrepreneur. However, if there is clearly an existing need for assistance in this area, and if the need is not met - despite the fact that the business success of the investee depends upon it - then this must be evaluated as a serious deficit on the part of investors.

In the remaining areas (**organization/personnel, R&D, production**), support was at a very low level; NTBFs also turned down assistance in R&D and production questions. Founders expressed virtually no need for help in these areas. The same applied to a great extent to organizational issues. These often tend to arise only after a certain order of magnitude is attained, which most of the firms in the survey had not yet reached.

NTBFs definitely hoped for assistance in **operative questions** and questions relating to the **day-to-day running of the firm**. However, in accordance with their servicing strategy, most investors do not offer support in these areas; they point to the unfavourable cost/benefit relationship.

The statements of the 42 NTBFs interviewed in the context of the BJTU pilot scheme generally reveal **substantial incongruencies between the need for support and the support received**. These can be attributed in part to the fact that the needs and expectations of the founders do not coincide with the offer of the investors. The greatest deficits emerge in the areas of marketing/selling and business management, i.e. exactly at those points which are ultimately decisive for the success of a NTBF.

3.3   Assessment of Management Support

Whereas the 42 NTBFs interviewed described the amount of capital raised as sufficient, many of their statements on management support express dissatisfaction. Although the relations to the investor generally were mostly described as a partnership and as informal (in the positive sense), assessments of the quality of the support provided sometimes reflect a quite different picture.

Approximately half of the NTBFs experienced substantial deficits in the quality and scope of the support. It should, however, be borne in mind that the expectations of some of the interviewees regarding the capabilities of investors were too high. These expectations were raised to some extent by the investors themselves (for instance by advertising brochures and in initial contacts). Nevertheless, it would be an over-simplification to explain away the dissatisfaction of numerous founders with their investors as a consequence of high expectations. Section 3.2 showed discrepancies - in some cases substantial discrepancies - between the need for support and the support received. These discrepancies are particularly evident in the area of marketing, so essential to the success of the enterprise. However, in other areas, too (business strategies, crisis management), many investors were not active enough, according to assessments by their portfolio enterprises.

Despite the great need for this, more than half the NTBFs stated that their investor had not, so far, fulfilled one important support function: the **mediation of contacts**. At most, contacts were mediated to other investors, lawyers and tax advisers. By contrast, access to industrial cooperation partners, customers and sales partners was rarely set up. It is particularly obvious that only seed and venture capital companies performed any "networking" activities to speak of (Bygrave/Timmons 1992). With the other types of investors, virtually no contact mediation took place. A possible cause for this is the networking of the investors themselves, which often tends to be weak outside the finance sector.

In addition, more than half the NTBFs complained that their investors did not make themselves available as **critical sparring partners** in the discussion of strategic matters or day-to-day problems. This leads to the conclusion that many business investment companies only partially fulfil either the needs of these firms, or their own aims to do so. For instance, in a series of interviews with 33 capital investors active in the early-stage investment segment (Wupperfeld 1996), most investors interviewed - with the exception of the MBGs - emphasized the particularly intensive way in which they performed their role as sparring-partners.

One cause for support deficits, apart from low personnel capacities or a general policy of not providing comprehensive management support, is probably insufficient qualification and experience of the investors. This supposition is backed up by the following statements about the **qualification of venture managers** in the estimation of the founders interviewed: more than one quarter of founders described the venture managers assigned to them as unqualified, inexperienced or unprofessional (cf. Figure 9); however, 40 percent considered them to be qualified. The assessment of the remainder was ambivalent.

Figure 9:   **Quality of management support - assessment by NTBFs**

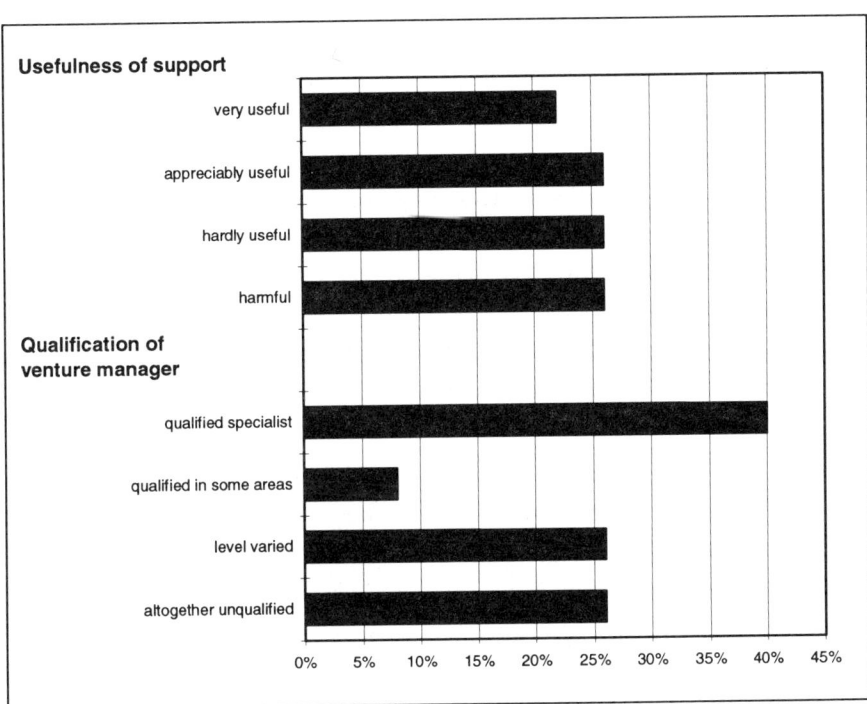

Estimations of the **usefulness of management support** are correspondingly divergent. The spectrum of answers ranged from: "Without the support, the firm would not be here today", "The support was really useful", "The business investment company was very committed" to: "The support was not useful in any way and was in fact harmful", "The investor was not really committed". Altogether, about equal proportions of interviewees stated that management support had been very useful, perceptibly useful, hardly or not at all useful, and negatively useful. The latter category means that from the viewpoint of the NTBF, the investor had actually done damage, for instance by insisting on strategies that were inadequate, by forcing rapid growth, by making erroneous decisions or by mediating unsuitable cooperations. These assessments do, however, include cases in which a conflict of interests arose between the investor and the NTBF.

Although the results obtained cannot be regarded as statistically significant due to the low figures involved, the following statements can be made on investor-related tendencies: independently of the type of investor, there are investors who provide qualified, useful support and ones whose support is unqualified. Having said this, the help given by seed and venture capital companies was often assessed as qualified and useful, whereas the support provided by credit institutions and business investment companies in the finance sector was frequently described as less qualified. The two latter types of investors were accused of a lack of commitment and of acting merely in their capacity as investors. Although MBGs do not generally offer (comprehensive) management support, cooperation with them was definitely assessed as positive. Some individual founders even emphasized the advantages of the silent partner's "hands-off" attitude. From this, it can be concluded - and some interviewees explicitly stated this - that NTBFs do not expect support services from the MBGs, since they perceive them as being promotion institutions only.

Another result of the survey emerges clearly if the assessments of the intensity, usefulness and deficits of support services are contrasted with the growth rates of the NTBFs. For this purpose, the firms were divided into two groups: those with a fast or medium rate of growth and those with a slow rate of growth (von Wichert-Nick/Kulicke 1994). This led to the following **significant result**: NTBFs with fast and medium growth rates received more intensive support, and evaluated its usefulness more positively, and its deficits less negatively, than did slowly-growing firms. It can be assumed that the founders of successful enterprises were altogether more satisfied and made more positive statements about their investors. The others, on the other hand, probably felt that the investors were partly to blame for their "lack of success" and expressed their disappointment in the interviews. However, the significant interrelation can be considered to substantiate the thesis that intensive and qualified support does in fact lead to faster growth and greater success for the NTBF.

4.   Conclusions

The TOU/ABL pilot scheme showed that during their start-up process, the needs of NTBFs for support vary greatly in extent and differ in intensity. Needs are oriented first and foremost towards strategic questions, but also towards operative aspects in marketing and selling. Neither private nor public advisory institutions are generally able to fill the role of "ideal advisors", dealing with the full range of specific - and usually very high - demands in their full breadth. Moreover, the financial means of NTBFs are not adequate to pay for qualified consultants. Many founders also have a sceptical attitude towards bringing in a consultant.

The approach adopted in the BJTU pilot scheme, in which the provision of management support for NTBFs was shifted to the investors, was only partially successful. In line with their general business policy, a proportion of the business investment companies and credit institutions taking up the promotion offer of the pilot scheme do not provide comprehensive advisory support and thus do not have any substantial capacities in investment management. However, even in the case of investors who define their offer as including the provision of capital plus management support, there are sometimes serious disparities between the need for advisory services and the support given. For this reason, only a part of the founders interviewed were satisfied with the non-financial assistance they received from their investors, assessing it as useful and efficient.

The need for consulting in NTBFs can also be ascribed to the fact that German universities, especially in technical disciplines scarcely deal at all with the aspects of innovation management or marketing and selling. The results relating to support services provided by investors also underline that here, too, substantial learning processes are necessary.

5. Bibliography

Baaken, T. (1989): Bewertung technologieorientierter Unternehmensgründungen. Kriterien und Methoden zur Bewertung von Gründerpersönlichkeit, Technologie und Markt für Banken und Venture-Capital-Gesellschaften sowie für die staatliche Wirtschafts- und Technologieförderung. Berlin.

Bayer, K. (1990): Beratung und Betreuung junger Technologieunternehmen. Erfahrungen aus dem Modellversuch TOU. Working paper. ISI: Karlsruhe.

BMFT (Bundesministerium für Forschung und Technologie) (Publ.) (1989): Beteiligungskapital für junge Technologieunternehmen. Modellversuch 1989 bis 1994. Bonn.

Bräunling, G./Gerybadze, A./Mayer, M. (1989): Ziele, Instrumente und Entwicklungsmöglichkeiten des Modellversuchs "Beteiligungskapital für junge Technologieunternehmen" (BJTU). Working paper. ISI: Karlsruhe.

Bygrave, W.D./Timmons, J.A. (1992): Venture Capital at the Crossroads. Boston.

Cohen, R. (1988): Venture Capital - More Than Money. In: Venture Economics (Publ.): Guide to European Venture Capital Sources. 2nd Edition. London, 15-17.

Fredriksen, Ø./Klofsten, M./Landström, H./Olofsson, C./Wahlbin, C. (1987): Entrepreneur-Venture Capitalist Relations: The Entrepreneurs' View. In: Churchill, N.C. et al. (Eds.): Frontiers of Entrepreneurship Research 1987. Proceedings of the Seventh Annual Babson College Entrepreneurship Research Conference. Wellesley, Mass., 251-265.

Gerybadze, A. assisted by Kulicke, M. (1990): Erfolgsbedingungen und -kriterien für junge Unternehmen, insbesondere für technologieorientierte Unternehmensgründungen. Working paper. ISI: Karlsruhe.

Gerybadze, A. (1988): Organizational Life-Cycle of New Technology-based Firms. In: Anglo-German Foundation (Publ.): New Technology-based Firms in Britain and Germany. London.

Gorman, M./Sahlmann, W.A. (1986): What do Venture Capitalists do? In: Ronstadt, R. et al. (Eds.): Frontiers of Entrepreneurship Research 1986. Proceedings of the Sixth Annual Babson College Entrepreneurship Research Conference. Wellesley, Mass., 414-436.

Hemer, J./Kulicke, M. (1995): Unternehmenskrisen in jungen Technologieunternehmen - Eine empirische Analyse der Krisenverläufe von im Modellversuch "Beteiligungskapital für junge Technologieunternehmen" (BJTU) begünstigten Unternehmen. Working paper. ISI: Karlsruhe.

Kulicke, M. u.a. (1993): Chancen und Risiken junger Technologieunternehmen. Ergebnisse des Modellversuchs "Förderung technologieorientierter Unternehmensgründungen". Heidelberg.

Kulicke, M. assisted by Gerybadze, A. (1990): Entwicklungsmuster technologieorientierter Unternehmensgründungen - Merkmale von Unternehmenstypen, die für Beteiligungs- und Kooperationspartner mit spezifischen Anforderungen hinsichtlich der zu erwartenden/möglichen Schwierigkeiten und den entsprechenden Unterstützungsleistungen verbunden sind. Working paper. ISI: Karlsruhe.

Kulicke, M. (1990): Enstehungsmuster junger Technologieunternehmen. Working paper. ISI: Karlsruhe.

Kulicke, M./Wupperfeld, U. in collaboration with Hemer, H./Traxel, H./von Wichert-Nick, D. (1996): Beteiligungskapital für junge Technologieunternehmen. Ergebnisse eines Modellversuchs. Heidelberg.

MacMillan, I./Kulow, D.M./Khoylian, R. (1986): Venture Capitalists' Involvement in their Investments: Extent and Performance. In: Kirchhoff, B.A. et al. (Eds.): Frontiers of Entrepreneurship Research 1988. Proceedings of the 1988 Entrepreneurship Research Conference. Wellesley, Mass., 303-323.

Mayer, M./Kulicke, M./Müller, R. (1986): Beiträge von Marktberatern zur Fundierung eines Unternehmenskonzeptes im Rahmen des Modellversuchs "Förderung technologieorientierter Unternehmensgründungen" (TOU) des BMFT. Working paper. ISI: Karlsruhe.

Picot, A./Laub, U.D./Schneider D. (1989): Innovative Unternehmensgründungen. Eine ökonomisch-empirische Analyse. Berlin.

Rosenstein, J./Bruno, A.V./Bygrave, W.D./Taylor, N.T. (1989): Do Venture Capitalists on Boards of Portfolio Companies Add Values Besides Money? In: Brockhaus, R.H. et al. (Eds.): Frontiers of Entrepreneurship Research 1989, Proceedings of the Ninth Annual Babson College Entrepreneurship Research Conference. Wellesley, Mass., 216-229.

Rosenstein, J./Bruno, A.V./Bygrave, W.D./Taylor, N.T. (1987): How much do CEOs value the Advice of Venture Capitalists on their Boards? In: Churchill, N.C. et al. (Eds.): Frontiers of Entrepreneurship Research 1987. Proceedings of the Seventh Annual Babson College Entrepreneurship Research Conference. Wellesley, Mass., 238-250.

Sapienza, H.J./Timmons, J.A. (1989): Launching and Building Entrepreneurial Companies: Do the Venture Capitalists Add Value?, in: Brockhaus, R.H. et al. (Eds.): Frontiers of Entrepreneurship Research 1989. Proceedings of the Ninth Annual Babson College Entrepreneurship Research Conference Wellesley, Mass., 245-257.

Schröder, C. (1992): Strategien und Management von Beteiligungsgesellschaften: Ein Einblick in Organisationsstrukturen und Entscheidungsprozesse von institutionellen Eigenkapitalinvestoren. Baden-Baden.

Unterkofler, G. (1989): Erfolgsfaktoren innovativer Unternehmensgründungen. Ein gestaltungsorientierter Lösungsansatz betriebswirtschaftlicher Gründungsprobleme. Frankfurt am Main.

Von Wichert-Nick, D. assisted by Kulicke, M. (1994): Ökonomische Entwicklung und Unternehmensstrategien junger Technologieunternehmen. Ergebnisse einer Befragung von im Modellversuch "Beteiligungskapital für junge Technologieunternehmen" (BJTU) begünstigten Unternehmen. Working paper. ISI: Karlsruhe.

Wupperfeld, U. assisted by Hemer, J. (1995): Die Betreuung junger Technologieunternehmen durch ihre Beteiligungskapitalgeber - Empirische Untersuchung. Working paper. ISI: Karlsruhe.

Wupperfeld, U. (1996): Management und Rahmenbedingungen von Beteiligungs-gesellschaften auf dem deutschen Seed-Capital-Markt. Frankfurt am Main.

II.   The Financing of Technology-Based Firms

The Promotion of New Technology-Based Firms in Germany

Marianne Kulicke

1.   Overview of the Promotion Offer

Since the 1960s, and even more since the 1970s, the promotion of firm foundations in Germany has ranked high among the economic policy measures of the Federal Government and the governments of the German Länder. The quantitative expansion of this promotion aspect over more than twenty years served to combat the deficit in firm foundations at that time (negative balance of foundations to closures). The promotion measures initiated then were intended to have a lasting effect in stimulating founding activities in all sectors of trade and industry, thus producing a rejuvenation and expansion of the existing population of enterprises. Quantitatively speaking, the most important promotion programmes are the "Eigenkapitalhilfeprogramm" (Equity assistance programme, EKH) and the "ERP-Existenzgründungsprogramm" (European Recovery Programme loans on business set-ups). These fall within the area of competence of the Federal Ministry of Economics.

Since the early 1980s, the attention of science and technology (S&T) promotion by the Federal Ministry for Research and Technology, now Federal Ministry of Education, Science, Research and Technology (BMBF) has been directed towards new technology-based firms (NTBFs). This development was triggered by monitoring reports which emphasized the special role played by innovative new enterprises in the USA, particularly in Massachusetts and in the Silicon Valley. Other contributory causes were the discussion, arising at about the same time, on Europe's tendency to fall behind in the international technology competition, and contemporary efforts to force technology transfer from publicly-subsidized universities and research institutions.

Over the last few decades, an increase in the numbers of players and measures in S&T is observed in Germany (Meyer-Krahmer/Kuntze 1992); this also applies to the group of NTBFs. Specifically with this group in mind, classical technology and innovation policy has been enlarged by the use of new instruments, such as an advisory component to complement financial support; infrastructural measures like the creation of incubator and technology centres; and promotion involving business investment capital and integrating private investors (cf. various contributions in this reader). As well as the European Union (EU), the BMBF and the German Länder, regional public and semi-public institutions have also begun, since the early 1980s, to offer support to NTBFs.

Foundation activities in Germany are promoted primarily by measures that address individual enterprises (cf. for instance Kurz et al. 1990; Schmude 1995; Kulicke 1995). Measures address the financing side; they promote market entry and the stability and resilience of new firms. There are also programmes with an advisory content, designed to help solve the economic, technical, financial and organizational problems associated with starting-up and managing an enterprise. On the other hand, the general body of state rules and regulations influencing the operating of enterprises does not contain any built-in advantages specifically for new or small enterprises. These can only benefit from special regulations in individual areas (such as tax law, for instance).

At a **federal level**, there has been a dual emphasis in programmes promoting foundations over the last few years, with two substantially different target groups:

- **Promotion programmes for the foundation of businesses in trade and industry and the liberal professions**: in these programmes, the promotion instruments used have primarily been long-term or equity-type loans. Within the Federal Government's financing support schemes for small and medium-sized enterprises (SMEs), these are classified with the general loan programmes for the financing of real investments. The funds for these programmes are drawn mainly from the ERP Special Fund or are own funds of the Reconstruction Loan Corporation (Kreditanstalt für Wiederaufbau, KfW) and the German Equalization Bank (Deutsche Ausgleichsbank, DtA). Since these actions address a broad target group, limiting the administrative complexity - also with regard to individual instances - plays a role in fixing the conditions for these programmes. For the same reason, there is no sectorally-specific differentiation of promotion conditions.

- **Promotion programmes for technology-based foundations and new technology-based firms**: these were the pilot schemes "Promotion of New-Technology-Based Firms" in the old Länder, TOU/ABL, "Promotion of New Technology-Based Firms in the new Länder", TOU/NBL and "Business Investment Capital for New Technology-Based Firms", BJTU. These were planned as temporary pilot schemes with the aim of gathering information about the specific support requirements of foundations of this kind (cf. Bräunling et al. 1989; Kulicke et al. 1993; Bräunling et al. 1995; Kulicke/Wupperfeld 1996). Another aim of the promoter was a stronger involvement of private sector investors in the financing of NTBFs. Compared with the promotion programmes for foundations in trade, industry and the liberal professions, these actions used a much more differentiated repertoire of measures. As well as subsidies, loans and credit guarantees, the emphasis in the past few years has been mainly on investment capital to improve the equity capital base. As well as real investments, it has also been possible to support operating costs (above all personnel costs), which account for a substantial proportion of expenditure in innovation projects.

All three pilot schemes have now expired according to plan.

In addition to such schemes, financing assistance for SMEs is also provided by the Federal Government and the Länder; this assistance is also available for new foundations, but does not have any special conditions for new firms (Kulicke 1995).

Table 17 shows Federal Government programmes for new foundations in trade, industry and the liberal professions, and for NTBFs. It is clear that the EKH programme and the ERP loan programme place emphasis on the promotion of investments by supported enterprises. In their start-up phase, however, the expenses that NTBFs have to finance are mainly personnel costs and operating costs (Kulicke et al. 1993). Investments only account for a relatively small proportion of their total expenditure at this stage (of the order of 20 %). Since it is usually necessary to approach several different sources in order to cover their requirements for capital, many NTBFs make use of the EKH or ERP loan programme to finance their own investments. Both measures thus have a complementary function for firms of this type.

The Länder also promote firm foundations, mainly by long-term loans and subsidies. These are primarily oriented towards business foundations with a smaller capital requirement than NTBFs, and are intended mainly for the financing of investments and the initial equipping of the enterprise (furnitures and fixtures). They sometimes include sector-specific promotion conditions which reflect the regional

economic structure. So far, only Baden-Württemberg has offered a promotion programme exclusively for NTBFs.

All evaluations of public programmes for the promotion of foundations (e.g. Huns-
diek/May-Strobl 1987; Kurz et al. 1990; Schmude 1995) indicate the very important
role played in the financing of foundations by public promotion funding. In particular it can be ascertained that its relative importance has increase continuously since
the early 1980s, causing credit financing by banks to lose ground.

Taken in a broader sense, promotion offers for NTBFs also include infrastructural
support, in the form of

- technology and incubator centres,
- technology transfer units at universities, senior technical colleges and non-
  university research institutions,
- institutions with the purpose of spurring technology transfer and consulting in
  aspects relevant to technology and innovation,
- offers of consulting services for SMEs by chambers of industry and commerce,
  on questions relating to the transfer of information and personnel, innovation
  consulting, qualification and further training measures in new technologies, as
  well as
- new forms of inter-community cooperation (usually under the rubric of "technology regions").

Only some of these offers are oriented primarily towards NTBFs. Infrastructural
assistance of this kind is examined more closely in paragraph 3.

Table 17: Federal programmes for the promotion of foundations (Key: S=subsidy, L=loan) (Status: October 1995)

Promotion programme	Content of promotion	Promotion instruments and target group
Measures applying throughout Germany		
• **EKH, equity assistance programme**	Risk-bearing funds to strengthen the equity capital base in the founding of an enterprise or a self-employed business, acquisition of an enterprise or permanent establishment also through privatisation, strengthening investments and follow-up investments within 3 years of first receiving support	L, natural persons under 50 founding an enterprise in small and medium industry or in the liberal professions, or wishing to participate as an acive partner in an activity of this kind
• **ERP loans on business set-ups**	Setting-up and acquiring businesses, and related investments up to 3 years after opening the business; takeover of partnership with managerial function; acquiring a first merchandise inventory or initial office equipment	L, Business foundations in trade and industry and in the eligible area also members of the liberal professions, with the exception of medical and paramedical professions
• **DtA business start-up programme**	Same promotion content as in the ERP loan programme for the promotion of foundations, complemented by foundation-related investments during the start-up phase of a business	L, Business foundations in trade and industry and the liberal professions
Measures applying only in new Länder and former East Berlin		
• **Complementary equity programme (EKE)**	So-called "soft" investments, i.e. non-material investments such as: the costs usual in the sector for opening up the market, with foreseeable longer-term tie-up of capital, costs for product development and market entry, project related training measures and costs for temporary management support. If the promotion possibilities of the equity assistance programme are exhausted, investments in the firm's fixed assets and merchandise inventory can be included	L, independent enterprises in small and medium trade and industry, with a turnover up to 250 million DM, particularly also new businesses
• **Pilot scheme "New Technology-Based Firms" in the new Länder and East Berlin**	Expenditure on elaboration of business concept and on performance of R&D, as well as market entry and preparation, for production of innovative products, processes and services	S, L, Persons wishing to found a technology-based business; technology-based enterprises not more than 2 years old, with not more than 10 employees, whose founders hold at least 51 % of the business capital

2. Promotion Instruments of the BMBF's Pilot Schemes for New Technology-Based Firms

2.1 The Pilot Schemes TOU/ABL and TOU/NBL

When the pilot scheme "Promotion of New Technology-Based Firms" (TOU) was started in the old Länder in 1983, little information was available on the problems of NTBFs in Germany. The main barriers were thought to be the financing problems and know-how deficits (in non-technical aspects) of firm founders. It was the intention of the promoter to gain information about

- how a climate conducive to firm foundations can be created for NTBFs,

- to what extent these firms contribute to enlarging the innovative technological supply,

- how their start-up and growth conditions can be improved, and

- how to stimulate the supply of risk-bearing capital for firms of this kind.

From these considerations, the aim was derived of providing consulting - also by private consultants - in addition to capital. There was an expectation that, in this way, the consulting offer of both private and public advisors would be built up and expanded.

With the TOU pilot scheme, the BMBF was breaking **new ground** in various ways (cf. Kulicke et al. 1993):

- **The target group**: this included newly-founded technology-based firms, existing manufacturing firms (up to three years old, not more than 10 employees), and service enterprises which were attempting, with their innovation project, to break into the manufacturing sector (up to six years old, not more than 20 employees).

- **There was no "blanket" promotion; definite stipulations were made for eligibility**: the firms eligible to apply were NTBFs registered in any of six promotion regions or in 15 technology and incubator centres, or active in the areas of biotechnology and microelectronics, or those in which an investment company was participating.

- **Promotion instruments**: these consisted of a financial component (non-repayable subsidies, credit guarantees) and an advisory component. The BMBF financed eight technology advisory offices (TBS) which, besides administering the scheme, also offered advice on starting-up the enterprise.

- **Support for activities preceding and following the R&D phase itself**: this includes elaboration of the business concept, as well as setting up production and market entry. The maximum amount and rate of funding varied with the promotion phase. The BMBF acted as guarantor for house bank credits required to finance market entry and setting up production.

- **Administration of the scheme**: administrative tasks were divided between the technology advisory offices and the promoter. The BMBF ultimately based its decisions on the recommendations of the advisory offices.

- **Cooperation with investors**: this related to promotion phase III (failure guarantees for house bank credits) and the access model of venture capital (eligibility of NTBFs with a participating investment company).

The TOU/ABL pilot scheme enlarged the repertoire of promotion instruments for technology transfer and was intended as a programme to stimulate and induce learning in technology policy. It was based on a linear sequence of foundation preparations, R&D phase and marketing phase. The three promotion phases were not allowed to overlap at all in time, and only slight overlaps in content were allowed. The promotion guidelines fixed clearly-defined stipulations for eligibility with regard to content and expenditure. Regarding the latter, beneficiaries very often had to submit individual documentary proof. The processes of application and administration were altogether very time-consuming. A year or more usually elapsed between the initial contact with a technology advisory office and a decision on acceptance for support. The work of the technology advisory offices was also very strongly characterized by involvement in administrative tasks relating to the promotion measures.

From 1983 to the end of 1988, a total of 319 new technology-based firms received, under the TOU/ABL pilot scheme, subsidies amounting to approximately 240 million DM for development work (promotion phase II). For 54 enterprises in promotion phase II and 97 in phase III (market entry and setting up production), the Federal Government guaranteed bank credits from house banks amounting to 5.3 and 91 million DM respectively. 258 (potential) firm foundations received subsidies totalling 7.6 million DM for activities relating to the elaboration of their business plan.

In May 1990, a new version of the TOU pilot scheme "New Technology-Based Firms in the new Länder", TOU/NBL, was started with the purpose of supporting the start-up of new technology-based firms in previous East Germany and in East Berlin. The instruments used in this new scheme were largely the same, but it contained modifications in the rates and maximum amounts of funding, in the administrative procedures and advisory services. Subsidies for the "business plan" in the foundation phase were fixed at 75 percent of a firm's total expenditure (with a ceiling of 45 thousand DM). Until April 1992, the R&D phase was supported by subsidies up to a level of 85 percent (maximum amount per firm: 850 thousand DM), and subsequently up to a level of 80 percent and 800 thousand DM per firm. For market entry and setting up production, project-related personal loans from the DtA (German Equalization Bank) of up to 500 thousand DM per firm could be made to partners actively involved in management. The consulting of the firms was entrusted to the VDI/VDE-Technologiezentrum Informationstechnik GmbH in Teltow, together with the Projektträger Biologie, Energie, Ökologie (BEO) affiliated to the Jülich Research Centre (Berlin office). The emphasis of the advisory component was on supporting development of the business concept, elaboration of marketing strategies and their realization, and the acquisition of capital for market entry. This promotion measure did not include any regional or sectoral limitations regarding the eligibility of newly-founded firms. Access to the scheme ended, as planned, at the end of 1995. Altogether, up to the end of 1995, 280 foundations were supported to the extent of 195.7 million DM. The follow-on measure, FUTOUR "Promotion and support of New Technology-Based Firms in Selected Regions" began in January 1997. This scheme combines subsidies, investments and founders' own funds, thus taking account of experiences gathered in the BMBF's previous pilot schemes, which indicated that a combined form of support is most appropriate for the promotion of NTBFs in the new Länder.

2.2   The BJTU Pilot Scheme

It did not prove possible, in order to avoid the situation of long-term subsidies, to transfer the promotion instruments used in the TOU/ABL pilot scheme to other responsible bodies, in particular the Länder - due to the financial circumstances and the small annual numbers of technology-based foundations in many of the Länder. Rather, the specific development requirements of this type of foundation made it desirable to use a new promotion measure aiming more directly at the activation of market forces, so that R&D-intensive and innovative technology-based firms would continue to arise in Germany in the future.

At the end of the 1980s, however, only a meagre supply of business investment capital was available for the early development phases of technology-based firms. As a consequence of negative experiences, venture capital companies, business investment companies in the finance sector - interested mainly in current income - and MBGs (Mittelständische Beteiligungsgesellschaften, investment companies in Germany formed by the Länder with the purpose of supporting small and medium industry) often limited their stake in NTBFs to a small proportion of their total investment volume, or excluded these firms generally as a group. The reasons for this attitude lay in the possibility of lucrative alternative investments in larger, well-established firms; the assessment of investment risks as being very high in new foundations (cf. Ruhnka/Young 1991); the limited internal resources of investment management; and the lack of re-financing options.

Use of the financing instrument of "investment capital" to attain the innovation policy aims had to be oriented towards the mechanisms of the investment capital market. Only in this way would it be possible to enhance its acceptance by both the supply and the demand side, and extend the market in the direction of NTBFs. By involving non-public investors in the support of NTBFs, the promoter was largely breaking new ground. Thus another promotion programme in the form of a pilot scheme was planned. It had the character of a **pilot scheme** (cf. BMFT 1989) in the sense that

- for the first time, it offered incentives to private sector suppliers of financing and consulting to invest much more extensively than before in NTBFs,

- it provided for different promotion conditions by including two different access models, in order to address a broad target group of investors, while at the same time ascertaining where the barriers to investments in NTBFs lay,

- it had the purpose of examining whether, and to what extent, the financing of such firms can take place as a result of market forces,

- it was intended to ascertain whether here, too, there is a long-term task for the public sector, and if so in what capacity.

Although it had the same target group, the BJTU pilot scheme differed from its predecessor in the following ways:

- **The form of financial assistance:** instead of direct subsidies to NTBFs, incentives were offered to investors of business capital (refinancing, deficiency guarantees, co-financing), to induce them to finance firms of this kind.

- **Consulting and advisory services:** this aspect was the responsibility of the investors, who also bore the relevant costs. In some individual instances, (profit-oriented) investors agreed with the recipients on a fixed charge or cost-related payment for their services. However, most investors cover their expenses by their income from investment or their gains on disposal.

- **The concept of two access models:** in order to appeal to a wide circle of business capital investors with specific restrictions in their commitments to NTBFs, the BJTU pilot scheme offered different promotion conditions (the co-investment and refinancing models, see Figiure 10). However, the access models did not differ with regard to the types of firms benefiting from them, but only with regard to the way in which the investments were supported. The refinancing model primarily addressed the barrier of "deficits in refinancing options when investing in NTBFs". This model was carried out by the Reconstruction Loan Corporation (KfW). The co-investment model was designed to combat the barrier of "high risks associated with NTBFs" and was carried out by the Technologie-Beteiligungsgesellschaft (tbg), a daughter of the German Equalization Bank (DtA).

Figure 10: Points of application of the two access models designed to combat barriers to investing in NTBFs

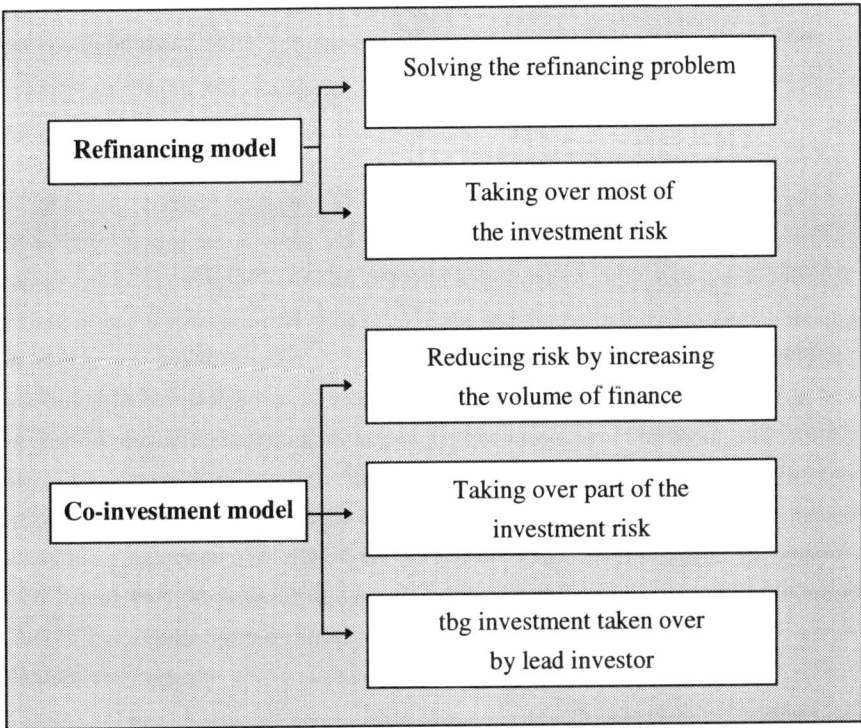

- **The type of innovation expenditure financed with the invested capital**: the BJTU pilot scheme did not insist on a precise stipulation of the types of expenditure eligible for support, the proportions of types of expenditure in total expenditure, etc. Nor was there a stipulated promotion level, i.e. a fixed share of the supported investment in the total financing volume of the supported firm. No differentiation was made in the promotion conditions between investments in the seed phase and the start-up phase. Investments were also supported which included both financing situations at the same time.

- **Administration of the scheme**: this was the responsibility of the KfW and the tbg, on the basis of contracts concluded with these two institutions. Coordination of open issues took place in bilateral and trilateral discussions. The KfW and the tbg themselves had a substantial interest in the success of the supported investments: the KfW bore a 10 percent loss in the case of non-repayment of refinanced loans and received 40 percent of the profits resulting from the investment. These included both the profit-dependent and the non-profit-dependent current income and the gains at disposal. The tbg bore the same share in the case of non-repayment of its investment and received income from investment to cover the expenses of selection, servicing and administering its commitment and its share of financial loss in case of failure.

The total investment capital mobilized by the BJTU pilot scheme included the silent partnerships of the tbg and the investments of the lead investors in the co-investment model, as well as the re-financed investments in the refinancing model. This amounted altogether to approximately 314 million DM of business investment capital and equity-capital-type loans for 336 NTBFs. However, the actual financing effect was manifested not only in the directly-supported investments, but also in the extent to which the supported firms were then able to open up other sources of capital. The supported investment capital was an important prerequisite for this: not only did it improve the equity capital base, and thus the creditworthiness of the founder; the commitment of one investor was regarded by other financiers as a positive signal for the success potential of the NTBF. When applying for public promotion funding, the invested capital was used to finance the own funds of the enterprise. As well as receiving investments and profit participating loans amounting to 327 million DM, all 336 NTBFs benefiting under the scheme were planning to use these funds to acquire:

- public promotion funding amounting to 101 million DM (subsidies and loans),

- loan capital amounting to 81 million DM (long-term and short-term credits), as well as

- 57 million DM own funds of founders and enterprises.

3. Infrastructural Assistance for New Technology-Based Firms

Since the mid-1980s, there have been numerous initiatives and measures at a regional and communal level, considerably expanding the previous range of tasks of regional economic promotion (cf. Ridinger/Steinröx 1995). These initiatives are intended to increase the innovativeness and competitiveness of firms located in the region ("strengthening the endogenous innovation potential") and build up the profile of the region in the competition with other communities to attract new businesses to locate there (cf. also the chapter on innovative regional development concepts in this reader).

Over the last ten years most German Länder have participated, together with communities, credit institutions and other private and public organizations, in setting up and running technology and incubator centres (cf. Koschatzky 1996; Sternberg et al. 1996). In March 1995, about 125 of these centres were regular members of the "Arbeitsgemeinschaft Deutscher Technologie- und Gründerzentren e.V." (the ADT, the society of German technology and incubator centres) (cf. ADT 1995). Newly-founded or small technology-based firms constitute the main target group for these centres, but they also target larger enterprises, in order to make the location attractive to other units with R&D tasks and business units with high value-added (cf. also the contribution on technology and incubator centres in this reader). In a few regions (e.g. Aachen, Cologne) special service centres have been set up to complement the offer of the technology centres.

In the old Länder, the founding of technology centres peaked in 1985 and 1986, with 23 and 20 new centres respectively being set up in these two years. Without a realistic estimate being made beforehand of the potential for NTBFs and young technology-based firms, these centres were set up not only in large cities, but also in small and medium-sized urban communities. However, in a considerable number of the centres, real demand did not match the over-optimistic target values. Thus over the years, these centres have developed into normal industrial estates (cf. n. a. 1988).

In the new Länder, too, technology and incubator centres were set up in many cities in the early 1990s. This occurred following the creation of an innovation service infrastructure for new and small technology-based firms (e.g. agencies for the promotion of technology and innovation, innovation advisory offices in chambers of industry and commerce, consulting establishments of the German Productivity Centre (Rationalisierungskuratorium der Deutschen Wirtschaft, RKW). They were financially supported to begin with by the BMBF, and later by the Länder governments (cf. Pleschak 1995). At that time, these technology and incubator centres helped NTBFs that were then in the process of formation, and would otherwise have had difficulty in finding and financing suitable rented premises during the period of transition to a market economy. In addition, they offer a wide range of consulting services relating to the problems of starting-up a business. Furthermore, they are regarded as important instruments and crystallization points for the development of an innovative SME industrial structure within a region.

Apart from these technology centres, over the last few decades other technology transfer institutions have grown up in universities, senior technical colleges and non-university research establishments in almost all cities in Germany (cf. for instance, BMFT 1993). In addition, supraregional establishments have been created, i.e. institutions based on individual Länder or sub-regions, with the purpose of spurring technology transfer and consulting in questions relevant to technology and innovation (e.g. ZENIT in North Rhine-Westphalia, TVA in Berlin, OTTI in East Bavaria, HIT in Hamburg, Bayern Innovativ in Bavaria) (see e.g. BMBF 1996). Most chambers of industry and commerce are also expanding their offer to include assistance for small and medium-sized businesses: advice on questions of information transfer and personnel transfer, innovation consulting, qualification and further training measures in new technologies. These are not specifically oriented towards newly-founded enterprises, but are aimed at small and medium-sized technology-based firms in general and at enterprises only sporadically performing R&D.

Infrastructural assistance for NTBFs also includes **new forms of inter-community cooperation** (e.g. the technology regions of Karlsruhe, Dortmund or Aachen). These aim to avoid inefficient competition between neighbouring cities for large enterprises intending to locate in the region, by generally seeking to create a local innovation network and a climate conducive to cooperation and innovation (see also the contribution on innovation networks in this reader). In their efforts to attract firms to locate, technology regions emphasize the presence of new or small technology-based firms in technology and incubator centres as possible suppliers and customers, or as a potential source of qualified personnel, thus constituting a locational advantage of the region.

As a rule, the forms of infrastructural support outlined above initially had their point of origin or their main focus in the support of NTBFs, but very soon acquired a broader group of addressees. The causes for this development were, on the one hand, a limited regional potential of NTBFs; on the other hand, a corresponding demand, which soon became apparent, from small and medium-sized (technology-based) firms. All in all, these infrastructural support projects led to an enhanced awareness, both in the public and the private sector, of the chances and problems of innovative enterprises.

4.   Conclusions

Following the **extensive changes made in its repertoire of instruments for the promotion** of NTBFs by the Federal Government in mid-1989 in the BJTU pilot scheme (business investment capital instead of subsidies and credit guarantees), in the last few years there have been several new approaches, also by the Länder governments (cf. for instance Kulicke 1995), to the promotion of small technology-based firms, also based on the instrument of investment financing. These either explicitly relate to the support of innovative new or small enterprises, or are designed exclusively for this target group. Two types of measures can be distinguished at the level of the Länder governments: one group aims to compensate what it considers as a lack of commitment on the part of private investment companies with regard to the financing of young and small technology-based firms, by making public funding available to enterprises. The other group (Bavaria, Baden-Württemberg and North Rhine-Westphalia) provide additional incentives (refinancing, risk reduction, co-financing) for investors in the private sector to participate more intensively in this segment of the economy.

All three of the Federal Government's pilot schemes for NTBFs have now expired. The follow-on measure, FUTOUR, was introduced in the new Länder in January 1997. From 1995, the repertoire of promotion measures used in the BJTU was incorporated - with modifications - into the promotion programme "Business Investment Capital for Small Technology-Based Firms", BTU, which addresses small and - in the new Länder - medium-sized technology-based firms. Thus at present there is no longer any nationwide promotion programme in Germany which is exclusively designed for NTBFs, but the present promotion programme, BTU, will be continued to the end of the millennium.

5.   Bibliography

ADT e.V. (Arbeitsgemeinschaft Deutscher Technologie- und Gründerzentren) (Publ.) (1995): Verzeichnis der Mitglieder. März 1995. Berlin.

BMBF (Bundesministerium für Bildung, Wissenschaft Forschung und Technologie) (1996): Bundesbericht Forschung 1996. Bonn.

BMFT (Bundesministerium für Forschung und Technologie) (Publ.) (1989): Beteiligungskapital für junge Technologieunternehmen. Modellversuch 1989 bis 1994. Bonn.

BMFT (Bundesministerium für Forschung und Technologie) (Publ.) (1993): Bundesbericht Forschung 1993. Bonn.

Bräunling, G./Gerybadze, A./Mayer, M. (1989): Ziele, Instrumente und Entwicklungsmöglichkeiten des Modellversuchs "Beteiligungskapitalmarkt für junge Technologieunternehmen" (BJTU). Working paper. ISI: Karlsruhe.

Bräunling, G./Pleschak, F./Sabisch, H. (1995): Ausgangslage, Ziele und Wirkungen des Modellversuchs "Technologieorientierte Unternehmensgründungen in den neuen Bundesländern" - erste Untersuchungsergebnisse. In: Holland, D./Kuhlmann S. (Eds.): Systemwandel und industrielle Innovation. Studien zum technologischen und industriellen Umbruch in den neuen Bundesländern. Heidelberg, 165-191.

Hunsdiek, D./May-Strobl, E. (1987): Gründungsfinanzierung durch den Staat - Fakten, Erfolge und Wirkungen. Schriften zur Mittelstandsforschung N.F. Nr. 17. Stuttgart.

Koschatzky, K. (1996): Relationship between Universities, Enterprises and "Technologieparks": The Role of the Länder. In: Gaetano, G. de/Logue, H. (Eds.): RTD Potential in the Mezzogiorno of Italy: The Role of Science Parks in a European Perspective. European Commission: Luxembourg, 51-55.

Kulicke, M. et al. (1993): Chancen und Risiken junger Technologieunternehmen. Ergebnisse des Modellversuchs "Förderung technologieorientierter Unternehmensgründungen" (TOU). Heidelberg.

Kulicke, M. (1995): Stellenwert junger Technologieunternehmen im Rahmen der Forschungs- und Technologiepolitik und Übertragung der Erkenntnisse aus ihrer Förderung auf neue Förderprogramme für kleine und mittlere Technologieunternehmen. Working paper. ISI: Karlsruhe.

Kulicke, M./Wupperfeld, U. (1996): Beteiligungskapital für junge Technologieunternehmen. Ergebnisse eines Modellversuchs. Heidelberg.

Kurz, R./Röger, W./Zarth, M. (1990): Existenzgründungshilfen von Bund und Ländern. Eine Wirkungsanalyse der Programme im Hinblick auf Wettbewerb, Produktivitätswachstum und Beschäftigung. Gutachten im Auftrag des Bundesministers für Wirtschaft. Tübingen.

Meyer-Krahmer, F./Kuntze, U. (1992): Bestandsaufnahme der Forschungs- und Technologiepolitik. In: Grimmer, K./Häusler, J./Kuhlmann, S./Simonis, G. (Eds.): Politische Techniksteuerung - Forschungsstand und Forschungsperspektiven. Opladen, 95-118.

n.a. (1988): Bald mehr als 100 Technologiezentren. In: Wissenschaft, Wirtschaft, Politik. Forschung und Innovation in Deutschland und Europa, No. 44-45, 4-7.

Pleschak, F. (1995): Technologiezentren in den neuen Bundesländern. Wissenschaftliche Analyse und Begleitung des Modellversuchs "Auf- und Ausbau von Technologiezentren in den neuen Bundesländern" des Bundesforschungsministeriums. Heidelberg.

Ridinger, R./Steinröx, M. (Eds.) (1995): Regionale Wirtschaftsförderung in der Praxis. Köln.

Ruhnka, J.C./Young, J.E. (1991): Some Hypothesis about Risk in Venture Capital Investing. In: Journal of Business Venturing (6), 115-133.

Schmude, J. (1994): Geförderte Unternehmensgründungen in Baden-Württemberg. Eine Analyse der regionalen Unterschiede des Existenzgründungsgeschehens am Beispiel des Eigenkapitalhilfe-Programms (1979 bis 1989). Stuttgart.

Sternberg, R./Behrendt, H./Seeger, H./Tamásy, C. (1996): Bilanz eines Booms. Wirkungsanalyse von Technologie- und Gründerzentren in Deutschland. Dortmund.

The Financing of New Technology-Based Firms

Marianne Kulicke

1.   Initial Situation

In the numerous discussions about securing Germany as a location for industry, it is repeatedly emphasized that technology-based firm foundations and young technology-based firms can make an especially important contribution to securing the international competitiveness of the economy. Since the beginning of the 1980s the Federal Ministry for Research and Technology, now the Federal Ministry of Education, Science, Research and Technology (BMBF), has promoted the formation of new technology-based firms (NTBFs) by three pilot schemes with various different promotion instruments and stipulations. These have differed substantially from the broad-based public programmes for founders in general (cf. for instance BMWi 1996). They are:

- "Promotion of New Technology-Based Firms" in the old Länder, TOU/ABL, which ran from April 1983 to December 1988 (last date of access);

- "Technology-Based Firms in the New Länder", TOU/NBL, from spring 1990 to December 1995;

- "Business Investment Capital for New Technology-Based Firms", BJTU, from mid-1989 to December 1994.

These promotion measures were planned as programmes to induce learning and stimulate activity. From them, a great deal of information was gained about the special financing and consulting requirements of NTBFs (Kulicke et al. 1993, Kulicke/Wupperfeld 1996, Pleschak/Rangnow 1995). This wealth of information flowed into the conception of new promotion actions, with target groups sometimes extending far beyond new foundations. The experiences from the TOU/ABL pilot scheme constituted the main basis for the design of the follow-on measure (the BJTU pilot scheme) and for the TOU/NBL pilot scheme. Knowledge from this promotion of NTBFs also flowed, for instance, into the concept for the "R&D Small Enterprise Loan Programme for the Application of New Technologies" and into the promotion programme "Business Investment Capital for Small Technology-Based Firms", BTU. Particularly from 1989 onwards the BMBF has aimed, with its BJTU pilot scheme, to stimulate the activity of private investors. This pilot scheme marked the introduction of a new instrument into the repertoire of the BMBF's science and technology policy measures: instead of direct funding for NTBFs, investors were offered incentives to inject business investment capital into the early development phases of firms of this type.

The Fraunhofer Institute for Systems and Innovation Research (ISI) was entrusted with the scientific monitoring of all three pilot schemes. The present contribution summarizes the results of these promotion measures with respect to the financing requirements of NTBFs and the existing financing offers.

2.   Financing Requirements of New Technology-Based Firms

The process of founding a technology-based firm usually originates with a more or less concrete product idea, and an initially rather vague idea of how to realize it. As these become progressively more concrete, the question very soon arises as to how much capital will be needed and the alternative options for meeting this need. However, the process of raising capital is beset by various imponderables resulting from the setting-up of a new business unit, and from the numerous and varied tasks and requirements involved in creating, out of the initial product idea, an innovative offer that is both feasible and viable (cf. also the chapters in Section I of this reader).

Typically, founders are not yet able to transfer into the new foundation concrete development results from their previous professional activities which would enable them to put a product on the market within a short period of time. Rather, they bring with them an abundance of know-how, concepts for their own products, experience and sometime also relations to potential customers, sales intermediaries, suppliers, research institutions, etc. These are the main constituents of their **"non-material start-up capital"**. The founder's own funds do not suffice to cover the financing of extensive development tasks for innovative products, processes or services, as well as market entry and the creation of manufacturing capacities. Founders have to raise additional capital.

The copious financing requirements become apparent if one reviews the types and structures of activities that are necessarily involved in the process of starting up NTBFs, and the financial inputs and outputs associated with them (see Table 18).

Figure 11: **Net capital requirements of the 336 NTBFs benefiting from the BJTU pilot scheme**

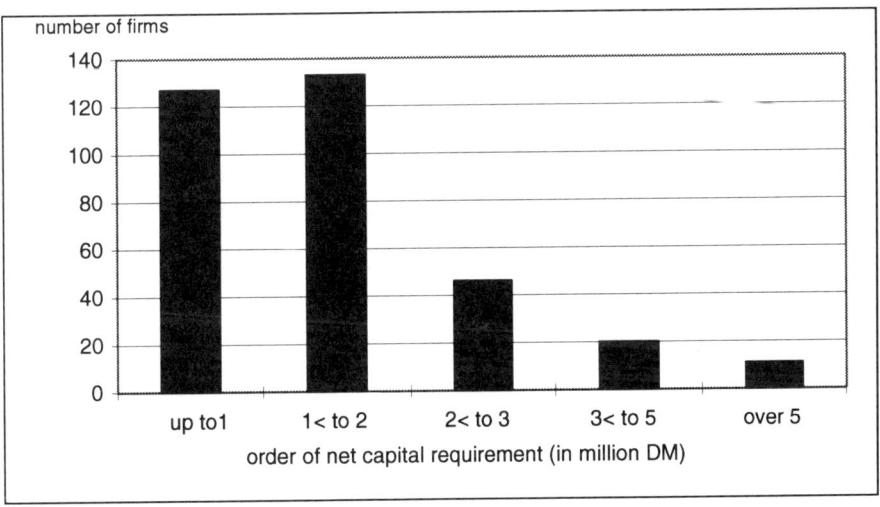

Figure 11 shows the size of the net capital requirement of the 336 NTBFs receiving subsidized business investment capital in the BJTU pilot scheme, which they used either to finance their development tasks or their expenditure on market entry and setting up production. This diagrammatic representation underlines the fact that both these stages in the start-up of NTBFs are associated with a great need for capital. However, the question of what sources can be tapped to meet this need depends not only on the general availability of capital, but also on the long-term aims of the founders with regard to pace of growth, company structure and their own position in the new firm.

Table 18:   **Activities in the process of starting-up NTBFs, and the financial inputs and outputs associated with them**

Phase	Main activities	Financial inputs and outputs
Elaboration of business plan	• Definition of the technical aims • Sounding out customer requirements • Evaluating market attractiveness and competition situation • Outlining the technical concept and the development steps • Estimating the need for financing • Raising loans, procuring equity capital • Establishing strategies, especially for marketing and selling	Input: small, particularly if all possibilities are used for externalizing these early-stage costs No output low net capital requirement
R&D	• Development tasks for innovative products or processes • Forming or expanding the R&D team • Cooperations in R&D • Continuous observation of the market • Initiating or intensifying contacts with market partners	High input: of the order of one or several million DM, depending on the type of project and available resources/ experience from previous development projects No output, or small output from this project via marketing of partial results High net capital requirement
Market entry	• Measures to make the new product or processes known (presentation at fairs, demonstrations, direct mailing, etc.) • Formal agreements with sales partners, suppliers and production subcontractors • Initiation or expansion of sale system (engaging or training personnel, setting up sales units etc.) • Authorization procedures • Testing by potential customers • Creating manufacturing capacities	High input: at least of the same order as in the R&D phase, sometimes much higher Output: first returns from market High net capital requirement

Phase	Main activities	Financial inputs and outputs
Market diffusion and growth	• Further expansion of sale system (particularly for foreign markets), • Building up (regular) clientele • If necessary, installing capital-intensive production plant for the (full-scale) manufacturing of the new development • Further development of the product range for additional applications/customer groups • Maintenance in product development	Input: high at first, depending on the volume of investments needed for production expansion, then substantially decreasing Output: high, in the case of success on the market Net capital requirement: definitely negative, i.e. profits are made, no external capital required

In interviews conducted by ISI with a great number of founders benefiting from the pilot schemes of the BMBF mentioned at the beginning of this chapter, it emerged that in most of NTBFs **raising capital** tended to be a **"trial-and-error" procedure** rather than a strategically-based involvement of partners in starting-up a new enterprise. There appear to be three main reasons for this:

1. **The founders' lack of experience** in the many different aspects of financing a firm, and in business matters in general. This also applies to the various financing offers available from public promotion funds and private sources. In this context, founders frequently complained of a great lack of transparency with regard to the conditions, financial implications, application stipulations and procedures.

2. **The absence of a "track record" and the high technical, market and general business risks** give rise to serious problems in assessing the time and the volume of financing necessary for starting-up the firm.

3. **"Language differences"** between founders - most of whom have a techno-scientific training and corresponding professional experience - and potential investors (banks, public funds, venture capital companies), who do not usually have enough relevant know-how to assess the business plan of a technology-based firm.

Founders generally complain of a **restrictive attitude of private investors** in the financing of their foundations. As well as the lack of data relating to past performance and the problems of evaluating technical aspects and market chances, mentioned above, the following reasons can be put forward for the restrictive behaviour of private investors:

- New enterprises are usually characterized by a **lack of real securities**; this is an aspect which plays a decisive role in reserved attitudes, especially with banks. Founders frequently pledge their own private assets, if they have any, as collateral for credits in the first few business years.

- In the case of credits or silent partnerships as a form of financing, the **participation in the chances and risks is asymmetrical** for investors in a new foundation: they bear a higher proportion of the loan loss risk, but this is not counterbalanced by correspondingly high returns (in the sense of risk premiums).

- Founders attach a great deal of importance to their **entrepreneurial independence** and are only prepared up to a point to take into the firm partners who have a say in decisions ("master in his own house" mentality). On the other hand, profit-oriented investors aim at direct, "hands-on" participation and thus envisage a co-partnership which will enable them to achieve high returns, to participate in business decisions and to combat erroneous developments.

In NTBFs, the question "How can I finance the foundation of my enterprise?" cannot be answered by simply listing the financing alternatives together with their pros and cons. Rather, the whole process of raising capital has to be considered in its entirety. The way in which the founder prepares and carries out this process decisively influences the extent to which an individual firm is really able to make use of the financing options that would be most favourable for it.

3. The Business Plan as a Starting Point for Capital Procurement

The starting-point for raising capital has to be a valid business plan; this has both a **planning** function and a **control function** (cf. Picot et al. 1989: 172, for the discussion of a "business development plan"). A business plan is necessary both for founders and for potential investors. From the founders' viewpoint, the written form in which their own ideas are fixed should provide an important base for valid strategic planning. Moreover, it can also clarify the risks associated with the foundation and the volume of resources required. For potential investors (investment companies, public promoters and banks) the business plan as a basis for investment firstly constitutes the most important **information instrument**. Secondly, it can be used as a basis for a first evaluation of the management's entrepreneurial skills: from the content and form of the plan, inferences can be made about founders' strategic thinking and their professionalism in cooperation with partners. The content of a plan of this kind is shown in Figure 12 (Geilinger 1991: 10f.).

Personal interviews with the founders of 42 new technology-based firms benefiting from the BJTU pilot scheme showed that some founders seriously underestimated the importance of a valid business plan (Kulicke/Wupperfeld 1996: 132f.): Just under one third of them only drew up a plan originally as a matter of form, in order to satisfy the requirements of investment companies. However the great majority, who drew up a plan at an early stage, (subsequently) regarded it as an important preparation exercise for planning the envisaged business development.

Figure 12:   Content of a business plan

- Object of the enterprise (offer to perform)
- Content and time plan for individual steps of start-up (development of products, market entry, creating production capacities, etc.)
- Aims of founders regarding growth, market position, their own position within the firm, ownership structure
- Experience of founders (technical, commercial and business knowledge, management and sales experience)
- Targeted markets (potential customers, competitors, barriers to market entry, targeted market segment and positioning relative to competitors, long- term development potential for the planned offer)
- Measures planned for opening up the market
- Assessment of capital requirements, plus ideas - or possibly already concrete activities - for covering them
- Description of risks and chances of the foundation, and
- Planned structuring of the organization

Figure 13 shows the frequency with which these individual components were contained in the business plans at the time of foundation, according to a written questionnaire answered by 118 new technology-based firms in the BJTU pilot scheme. (von Wichert-Nick/Kulicke 1994: 27ff.). Almost all of them had a cost plan; and from this the need for capital was derived. Closely linked with this was the planning of individual stages over time (76 %) for the realization of the innovation project and the creation of a stable enterprise. The frequency with which the product range was included (85 %) confirms that the foundation was generally based on a concrete product idea. On the other hand, there is a great deal of variation in the frequency with which business plans include indications on market aspects, although these are particularly relevant for investment companies: description of the market targeted (67 %), estimation of market potential (76 %), market share aimed at (46 %), positioning in the market (57 %). There is a similar situation with regard to the aims of the foundation, and the planned degree of structuring of individual functions: according to founders' statements, 81 percent of the business plans contained a description of growth targets, only 35 percent mentioned the short- and medium-term ownership structure aimed at in the new foundation, and 61 percent contained statements about the aims with regard to economic independence.

Figure 13: **Frequency of inclusion of individual components in the business plans of NTBFs benefiting from the BJTU pilot scheme (in %)**

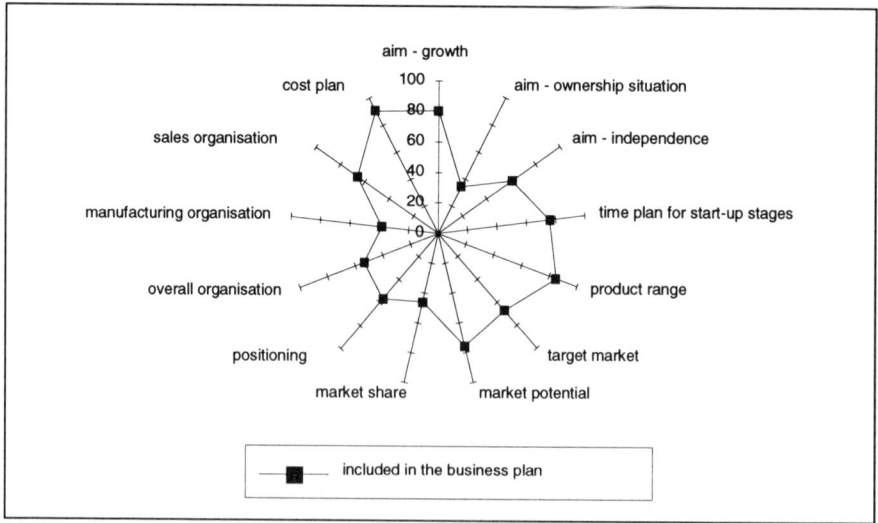

Investors of business capital in Germany frequently complain that new enterprises turning to them for capital have often not yet elaborated a valid business plan (Wupperfeld 1994: 90). The business strategies, and the planning derived from them, first have to be worked out in the course of the investment negotiations. However, for the viewpoint of business investment companies this is usually a very time-consuming process. As a consequence, their willingness to invest in enterprises of this kind is reduced.

4. Requirements for Investors in New Technology-Based Firms

From the particular features of the financing needs of NTBFs outlined above, the following requirements for investors in these firms can be derived:

1. The capital investment should be used first and foremost to **strengthen the equity capital base** and should have a **long-term character**, in order to minimize the strain on liquidity in the first few business years.

2. Since it is very difficult to estimate with accuracy the capital needed to finance the start-up of the firm, and founders often tend to underestimate the costs and the time required, an investor should have an adequate **potential for further financing** in order to meet unexpected financing needs.

3. Founders also have a **need for support in starting up the firm** which relates not only to the solution of financing problems but also to a wide range of other questions. Experience shows that this need centres around selling and marketing, as well as commercial and business questions (Kulicke/Wupperfeld 1996: 185ff.). It arises from the multiplicity of demanding tasks associated with starting up a totally new business unit. Founders do not usually possess the full range of experience and qualifications necessary for this; frequently at the beginning of the start-up process they still have substantial deficits in selling and commercial activities. This results in a need for support, which has to be met either by the investors themselves, or by the appropriate involvement of a third party.

Figure 14 summarizes the requirements made by NTBFs with regard to their investors.

Figure 14: **Requirements for investors resulting from the particular situation of NTBFs**

Financing particularities	Requirements: investors should:
Founder has limited own funds	strengthen the equity capital base
Numerous business risks	show flexibility in covering unforeseen financing needs provide financial and non-financial support, particularly if the original business aims are not achieved
High net capital requirement in the start-up years	make large amounts of capital available provide support in "tying up" the financing package avoid putting a strain on the firm's liquidity during the first few business years
Capital tied up for long period	provide long-term capital
Founders prejudiced against "hands-on" investors	make efforts to convince and give a realistic idea of their own performance profile

5. Availability and Limits of Alternative Financing Options

In order to cover their high capital requirements, NTBFs typically have to exploit several financing sources in the course of their start-up process. The initial steps are usually financed with own funds and small credits taken out under personal guarantees. If a need for further financing becomes apparent, or financing bottlenecks arise - which they generally do - then founders try to tap other sources of external financing. Here, four possibilities are generally open to technology-based firm foundations:

- Public promotion programmes (ones specially designed for new foundations, and ones intended for small and medium-sized firms in general),

- Credit financing (short- and long-term loans),

- Business investment capital (silent partnership and direct investments) from business investment companies,

- Business investment capital from informal investors.

Another possible source of financing is to start up a "service" business in the same field, using the profits to finance part of the expenditure on development work, setting up production or market entry of an innovative offer. In the following paragraphs, the four financing alternatives previously mentioned are discussed in more detail.

5.1 Public Promotion Programmes

Although there are numerous public programmes by the German Federal Government promoting individual enterprises, only a few of these come into question to finance the first start-up years of NTBFs (on this aspect, cf. the lists in von Freyend et al. (n.d.); Zeitschrift für das Gesamte Kreditwesen 1993). However, flanking these there are also promotion programmes of the individual Länder, which are only referred to briefly here. Of these, too, only a few go some way towards meeting the financing needs of NTBFs.

The great interest on the political side attracted by high-tech foundations has led, since 1983, to the three pilot schemes mentioned at the beginning of this chapter which were specially designed for the target group of NTBFs and made use of different sets of instruments. The first of these promotion actions was the pilot scheme TOU/ABL, which started in spring 1983 and was accessible until 31.12.1988. In the new Länder, NTBFs were promoted from spring 1990 to the end of 1995 (access period) under the TOU-NBL pilot scheme. A follow-on measure in the form of the modified promotion programme "Promotion and Support of New Technology-Based Firms", FUTOUR, has been running there since 1997. Until the end of 1994, a nationwide offer for NTBFs was provided by the pilot scheme BJTU. This was replaced at the beginning of 1995 by the promotion programme "Business Investment Capital for Small Technology-Based Firms", BTU, and was extended to include small (in the old Länder) and medium (in the new Länder) technology-based firms. Thus since the beginning of 1995, no nationwide promotion offer exclusively intended for NTBFs has been available in Germany.

However, NTBFs continue to constitute one of the main target groups of the BTU promotion programme. As in the previous BJTU pilot scheme, this promotion action addresses investors of business capital, offering them incentives to invest more risk-bearing capital in small and medium-sized technology-based firms. In the old Länder, the investments eligible for support are investments made by business investment companies, credit institutes, enterprises or private investors in small firms not more than ten years old. In the new Länder, this promotion can be extended to include medium-sized technology-based firms that are not more than ten years old.[1]

The BTU promotion programme has **two access models** with different promotion conditions. In the **re-financing model**, investments up to a ceiling of three million DM are re-financed at a fixed rate, provided the investor is bringing capital into small or medium-sized technology-based firms. If the investment fails, the Federal Government shoulders a 75 percent (in the old Länder) or 85 percent share (in the new Länder) of the loss of the re-financed investment. This access model is intended to appeal to investors with limited own funds, motivating them to take on high-risk commitments for the financing of innovation. This model is carried out by the Reconstruction Loan Corporation (KfW, Kreditanstalt für Wiederaufbau). In the **co-investment model** a daughter of the German Equalization Bank (DtA, Deutsche Ausgleichsbank), the tbg, acts as a silent partner, on the condition that a private investor (the so-called "lead investor") participates at least to the same extent and also provides support services for the recipient.

In both access models, the conditions are freely negotiated between investor and investee. The young firm in search of capital first has to identify a suitable investor and then convince him to invest. From the viewpoint of the firms, the main promotion effect of the BTU programme is a marked increase in the willingness of business investment companies to commit themselves, because the extensive cover greatly reduces the degree of risk attaching to the investment. Moreover, this scheme offers several different types of investors (particularly those that provide capital for the promotion of small and medium-sized firms, but also small seed capital companies) a first opportunity to commit themselves to investing in new and

---

[1] The delimitation of "small" and "medium-sized" firms is according to the EU guidelines: small enterprises are those with up to 50 employees and either a maximum annual turnover of ten million DM or a maximum annual balance sheet total of four million. DM. Medium-sized enterprises have a maximum workforce of 250 and either a maximum annual turnover of 40 million DM or a maximum annual balance sheet total of 20 million DM.

small technology-based firms. In this way, the circle of potential investors for young and small/medium technology-based firms is enlarged, both quantitatively and in terms of the forms of participation and investment conditions.

NTBFs are also eligible, however, for the **general promotion programmes of the Federal Government for the foundation of self-employed businesses**. These are the "Eigenkapitalhilfeprogramm" (equity assistance programme, EKH), the "ERP-Darlehen" (ERP loans on business set-ups, from the European Recovery Programme Special Fund) and the "DtA-Existenzgründungsprogramm" (DtA business start-up programme) of the German Equalization Bank, all of which are valid throughout Germany. These three offers address natural persons wishing to start a business in trade, industry or the liberal professions (the EKH programme also includes promotion for active partnerships). A common feature of the three programmes is that they all support business foundations by means of low-interest, long-term loans with a redemption-free start-up period (cf. Deutsche Ausgleichsbank 1994; Schmude 1994; Kurz et al. 1990).

Although the promotion ceilings for these three programmes have been substantially raised in the last few years (EKH: up to 700 thousand DM; ERP loans: up to 1 million DM; DtA business start-ups: up to 4 million DM), they all have a serious disadvantage from the viewpoint of NTBFs: they are intended primarily to finance the investments made by the supported businesses. Most newly-founded technology-based firms, on the other hand, first have to create their offer by extensive R&D and then bring it to the market-place. These are tasks associated with high personnel costs. Moreover, a promotion programme of this kind can, of course, only cover a part of the financing requirement in the initial start-up years. With the exception of the EKH programme, the Federal Government's general promotion programmes for self-employed businesses do not include any special conditions for technology-based foundations.

Since November 1995, the "Eigenkapitalergänzungsprogramm" (complementary equity programme, EKE) has been running in the new Länder. This programme includes loans for the financing of so-called "soft" - i.e. non-material - investments (costs usual in the sector for opening up the market, with foreseeable long-term tie-up of capital; product development and market entry costs; project-related training measures; costs for temporary management support). When the promotion possibilities of the EKH programme have been exhausted, investments in fixed assets and merchandise inventory can also be included. The target group for this programme is independent businesses in small and medium trade and industry with an annual turnover of up to 250 million DM, and also particularly self-employed business foundations.

Despite the disadvantages of the three first measures mentioned above, these promotion programmes of the Federal Government for business foundations in general do also fulfil an important function with regard to NTBFs. They usually make up a significant proportion of the total financing package of these firms, and are used for the financing of investments.

In principle, NTBFs are also able to participate in the general measures of the Federal Government for the promotion of R&D in individual small and medium-sized firms. However, with the exception of the BTU promotion programme outlined above, their suitability for the start-up financing of innovative foundations is very limited, due to the promotion instruments and conditions they incorporate. Programmes whose usefulness for NTBFs is limited include the BMBF's specialized S&T promotion programmes, the programme "Forschungskooperation in der mittelständischen Wirtschaft" (Research cooperation in small and medium industry) and the ERP innovation programme.

Apart from measures by the Federal Government, most of the Länder governments offer programmes for the promotion of industrial R&D and innovation, and the promotion of foundations. However, these are either in the area of self-employed foundation programmes and address new foundations in general, or they include the generality of small and medium-sized enterprises in a scheme to promote research and development. As a rule, they only cover a limited proportion of R&D expenditure (40 % at most) and they mostly do not include the very capital-intensive market entry phase.

5.2   Credit Financing

As various public promotion programmes are based on the loan principle, credit institutions fulfil an important function in transmitting funds from programmes of this kind. To some extent, they have to perform assessment and monitoring functions, paying out funds and controlling the use, and possibly also the repayment, of the funding.

Of course, banks do finance NTBFs with long-term loans via the usual credit financing mechanisms, but they will only do so if there are sufficient real securities, i.e. if the founders possess real property or other securities which they can use as collateral. For the founders, this represents a high degree of personal risk. Moreover, credit institutions usually arrange the credit for a firm in the form of a credit in current account, which is usually associated with a very high rate of interest. If there is insufficient collateral, the attitude of banks tends to be very restrictive - though from their point of view justifiably so - if they assess the risks involved in the foundation as being too high, and have problems in evaluating the technical and market aspects of the foundation project. These latter factors explain the complaints of founders, mentioned at the beginning of this chapter, about the reluctance of banks to provide credits to finance start-ups. However, founders also criticise the inadequacy of advice given by banks on financing alternatives. Founders attribute this partly to information deficits of individual advisors servicing the client firms; on the other hand, they also suspect that banks tend to give information primarily about programmes in which they themselves can be included as the transmitting credit institution.

However in the BJTU pilot scheme, and currently in the BTU promotion programme, a number of credit institutions also chose another way of supplying risk-bearing capital to NTBFs: **profit participation loans with waiver of priority**. Profit participation loans generally have a profit-dependent and a non-profit-dependent interest component, and the investor's rights of control are usually very limited (Weiss 1992: 233 und 218f.). There is no participation in losses and no responsibility in excess of the capital invested, i.e. in the case of bankruptcy, the entitlement of the party supplying the loan remains fully intact. It is also possible for the investor to transfer the loan to a third party without the assent of the recipient (this is one difference from a typical silent partnership). "Waiver of priority" means that in repayment, or in the case of bankruptcy of the recipient, this loan is only dealt with *after* the claims of the other creditors have been satisfied (otherwise, banks would not normally make loans in the form of "secondary" capital, but would insist on being repaid first). In the case of loss, secondary capital is liable directly after equity capital. If a company goes into liquidation, secondary capital is paid back before dealing with the claims of equity capital investors.

In the BJTU pilot scheme, approximately 70 million DM flowed into NTBFs in the form of profit participation loans. There were considerable regional differences, certainly due in part to a differing awareness of this option for the financing of enterprises by banks. The credit institutions that made loans of this type were primarily situated in North Rhine-Westphalia, especially in the Ruhr Valley area and in the region of Aachen. Of the 93 million DM of business investment capital flowing into NTBFs in North Rhine-Westphalia under the BJTU pilot scheme, as much as 50 million DM was in the form of profit participation loans from credit institutions.

Apart from this mode of supply of risk-bearing capital by banks, the limited willingness of credit institutions to finance NTBFs leads to the conclusion that credit financing can fulfil, at best, an auxiliary function in technology-based start-ups.

5.3   Business Investment Capital

Since the beginning of the 1980s, and especially since the start of the BJTU pilot scheme in 1989, the provision of risk-bearing capital for NTBFs by investment companies has increasingly come into the foreground in the promotion of enterprises in Germany. Since the end of the 1980s at Federal Government level, and at Länder level mainly since the mid 1990s, the trend has emerged of supporting new or small and medium technology-based firms primarily through business investment capital instead of subsidies. Nevertheless, this financing option is frequently not known. Founders in particular still have great information deficits regarding the chances and limitations of business investment capital, the various offers available from different types of investors, and the investment conditions.

One special feature of the German market for investment capital is the role of investment companies founded, with public support, as self-help organizations by industry, with the exclusive aim of promoting the economy. These associations, known as "Mittelständische Beteiligungsgesellschaften" (MBGs) provide equity capital for small and medium-sized firms and also for young enterprises in the form of silent partnerships. Their business policy is largely determined by public programmes offering re-financing and failure guarantees, since their own funds are very limited. The MBGs with the largest investment volume are situated in Bavaria, Baden-Württemberg, Hesse, Saxony and Thuringia. Although at the end of 1995 the publicly promoted investment companies in the "Bundesverband deutscher Kapital-beteiligungsgesellschaften - German Venture Capital Association e.V." (BVK), with investments of 952 million DM, only had a 16.9 percent share in the total volume of capital invested by all BVK members, their significance as investors in small and medium-sized enterprises is better indicated by their share in the number of investments: with 1,780 investments, this amounted to a share of 59.5 percent. In 1995 they showed by far the greatest increase in investment volume of the four specialist groups in the BVK (BVK 1996: 122)[2].

Private investment companies tend to engage mainly in the financing of established growth enterprises. In Germany, only a few profit-oriented companies of this kind have extensive commitments, or even specialize, in early-phase investments. However, the BJTU pilot scheme of the BMBF, which expired at the end of 1994, has contributed to a definite expansion, over the last few years, of the supply of venture capital to NTBFs by profit-oriented investors and also by the MBGs.

The BJTU pilot scheme showed that business investment capital has a substantial **leverage effect on the financing of NTBFs** as a whole. Roughly speaking, one DM of business investment capital mobilized just under two DM of other funds (Figure 15). This can be attributed to the significantly improved equity capital base after taking on business investment capital, and also to the fact that the commitment of one equity capital investor signals to other investors that a positive assessment of the new technology-based firm's development chances has been made by the former.

However, particularly when raising investment capital, it is very important that there should be a clear business plan containing founders' concrete proposals about the growth they are aiming at and their own intended position in the firm.

---

2 The other specialist groups in the BVK are: universal investment companies; seed-, start-up and growth investment companies; and UBGs (Unternehmensbeteiligungsgesellschaften).

Figure 15: **Planned means of financing the innovation projects of the 336 NTBFs benefiting from the BJTU pilot scheme**

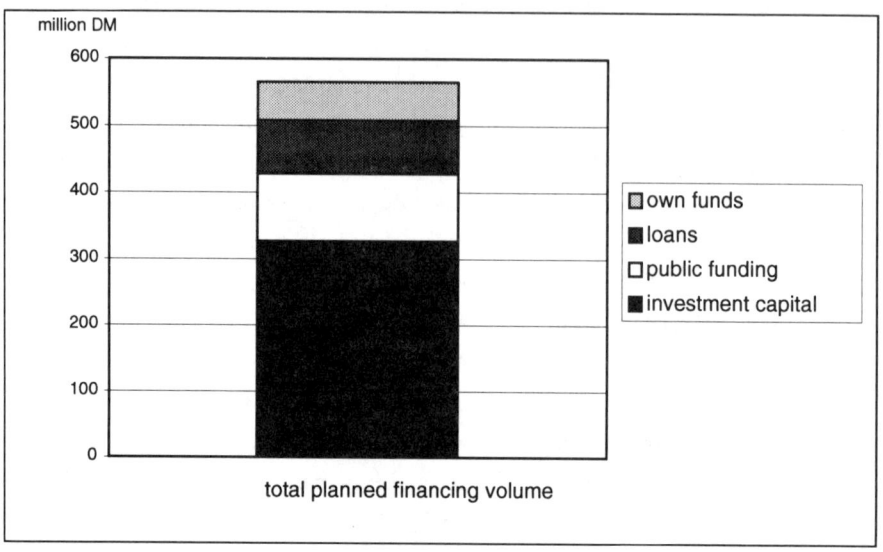

Dependent on the type of investor and the form of investment (mainly direct investment or silent partnership), the taking-on of investment capital is associated with different levels of payment but also with different support services. For their silent partnerships, MBGs typically agree on a non-profit-dependent interest component which at present is of the order of six percent. This is fixed for the whole period of participation. In addition, there is a profit-dependent component which ranges between two and three percent and depends on the size of the profits made by the firm. With profit-oriented investors, the rates of interest (in the case of silent partnerships) are usually considerably higher; however, there is also more scope here for manoeuvre by the recipient who may, for instance, make an agreement in which a relatively low fixed component is combined with more participation for the investor in the profits of the enterprise.

Most profit-oriented investment companies, however, aim at direct participation in the share capital, as they expect the value of the enterprise, and consequently the value of their share in it, to increase during the participation period, yielding high returns on their investment. Companies specializing in early-stage financing (seed capital companies) and venture capital companies describe their "product" as the provision of capital plus management support. Ideally, this support extends to all aspects related to starting up a firm.

Responses to a written questionnaire by 118 NTBFs receiving supported investment capital through the BJTU pilot scheme showed that, almost without exception, founders assessed the public promotion via business investment capital positively (von Wichert-Nick/Kulicke 1994). The aspects they mentioned most favourably were the improvement of the equity capital base, the possibility of realizing a very high proportion of the start-up financing with the business investment capital, and the much less time-consuming application and administrative procedures compared with public promotion funding.

Personal interviews with 42 entrepreneurs from NTBFs participating in the BJTU pilot scheme showed that the attitude of founders to business investment capital has definitely changed since the foundation (Kulicke/Wupperfeld 1996: 143f.). At the time of foundation, most of them (84 %) were pursuing the aim of keeping the majority interest within the circle of founders; only 16 percent of them would have accepted the position of a minority shareholder, if expansive growth of the enterprise required this. After the first few business years (average: 5.5 years after foundation) the picture had changed significantly: just under 60 percent of founders would also accept the position of a minority shareholder, if this would enable the firm to grow or would ensure its survival. This change in attitude was partially brought about by experiencing crises, which have to be regarded as symptomatic of the start-up years of NTBFs (Hemer/Kulicke 1995, cf. the contribution on the crises of NTBFs in this reader). However, positive experiences of cooperation with investors has also contributed to the change. It emerges particularly clearly from the responses of the 118 NTBFs that the firms whose investors were seed and venture capital companies mainly assess the promotion of enterprises through investment capital as positive.

By contrast, in the evaluation of support services received by NTBFs from their investors, serious deficits of German business investment companies also become apparent (Wupperfeld 1996). NTBFs' greatest need for support is in the area of marketing and selling. Thus, founders hope for support mainly in the opening up of (foreign) markets, the initiation of contacts with customers and sales partners, and the strategic orientation of marketing. However, most investment companies cannot meet these needs, since their investment managers rarely have an entrepreneurial background. Furthermore, they lack the necessary knowledge of the specific features of the markets of NTBFs. Investors are able to render important assistance, on the other hand, in acquiring further capital, building up an internal controlling system and solving other commercial problems.

5.4   Informal Investors

Whereas studies in the USA emphasize the great importance of informal investors (so-called Business Angels) for the start-up financing of foundations (meaning foundations in general, not NTBFs in particular), there are as yet no indications that Business Angels play any noteworthy role in Germany. In the USA, these informal investors are rich private persons, sometimes previous founders. They help to finance initial business activities, until the firm reaches a certain size and stage of development and can enter into negotiations with institutional investors. Contact to informal investors of this kind is usually in a regional context, or through previous professional contacts. Not only do Business Angels provide capital; they also - insofar as their business background enables them to do so - render assistance in the strategic orientation of the foundation, help to form contacts in the market and to other investors, and take on the role of "sparring partners" in discussions. Their contribution to capital is usually greatest at the beginning of the start-up process, falling to a small percentage when capital from formal investors flows into the successful new firm in further rounds of financing. There were also individual instances, among the NTBFs benefiting from the TOU/ABL and BJTU pilot schemes, of informal investors participating in the early start-up phase. Through longstanding contacts, these were usually able to make a realistic assessment of founders' business potential. Nevertheless, private investors definitely represent a source of capital which, as yet, is too little used - and also too rarely available - for new foundations in Germany.

6.   Summary

The start-up process of a new technology-based firm is associated with a great need for capital, necessitating the exploitation of various different sources of financing. Before commencing the process of raising capital, which is usually time-consuming, founders should elaborate a clear business plan for use as a planning and information instrument. Besides clarifying the volume of financing needed, and the uses to which it is to be put, this will also emphasize the requirements that the financing will have to satisfy (improvement of the equity capital base, long-term availability, not putting a strain on liquidity in the start-up years).

The difficulties, lamented by numerous founders, in finding sources of capital are attributable both to deficits on the part of the new firms, and to risk aversion on the part of investors caused by information deficits and by limited possibilities for return on investments.

However, no generally valid "magic recipe" can be given for the various financing alternatives resulting from public promotion programmes, bank credits and business investment capital regarding the combination or permutation in which they should be used for a technology-based firm foundation. Each form has its own specific advantages, drawbacks and limitations, so that their suitability can only be assessed in concrete individual instances. The need to cover the capital requirements of a technology-based foundation, which are generally high, usually means that founders have to avail themselves of several different offers.

Although many entrepreneurs complain of the great number and variety of public promotion measures, only a few of these are relevant for high-tech foundations. In the old Länder, there has not been any special promotion action exclusively for NTBFs since 1995. The measure that appears to be best suited to the financing requirements of NTBFs is the promotion programme "Beteiligungskapital für kleine Technologieunternehmen" (Business investment capital for small technology-based firms, BTU), as this measure enables large amounts of capital to be raised and the typical expenses associated with foundations to be covered. However, the general self-employed foundation programmes can also be considered as auxiliary promotion programmes, mainly for the financing of real investments. Usual bank credits, on the other hand, can only play a limited role, as they require the existence of real securities and put a great strain on the liquidity of a new foundation.

The financing requirements of NTBFs are best met by business investment capital in its "ideal" sense: long-term equity capital, participation of the financier in the risks and chances of a project, compatibility of the interests of investor and investee regarding the development of the firm, support of founders in financial and non-financial matters, etc. However, it is only since the beginning of the 1990s that a corresponding "financing culture" in the field of early-stage investments has slowly started to develop in Germany, now that the venture capital euphoria of the early 1980s, and the subsequent disillusionment regarding its possibilities and limitations in the prevailing framework conditions, have passed, giving way to a more realistic view.

7.   Bibliography

BMWi (Bundesministerium für Wirtschaft) (Publ.) (1996): Risikokapital für Existenzgründer und mittelständische Unternehmen. BMWi-Dokumentation. Bonn.

BVK (Bundesverband deutscher Kapitalbeteiligungsgesellschaften - German Venture Capital Association e.V.) (1996): Jahrbuch 1996.

DtA (Deutsche Ausgleichsbank) (Publ.) (1994): Richtlinie und Merkblatt der Deutschen Ausgleichsbank zum Eigenkapitalhilfe-Programm zur Förderung selbständiger Existenzen in den alten Bundesländern und Berlin (West). Stand Juni 1994. Bonn-Bad Godesberg.

Geilinger, U.W. (1991): Der Business-Plan. Eine praxisorientierte Anleitung zur Erstellung eines Business-Plans. 3rd Edition. Zürich.

Hemer, J./Kulicke, M. (1995): Unternehmenskrisen in jungen Technologieunternehmen - Eine empirische Analyse der Krisenverläufe von im Modellversuch "Beteiligungskapital für junge Technologieunternehmen" (BJTU) begünstigten Unternehmen. Working paper. ISI: Karlsruhe.

Kulicke, M. et al. (1993): Chancen und Risiken junger Technologieunternehmen. Ergebnisse des Modellversuchs "Förderung technologieorientierter Unternehmensgründungen" (TOU). Heidelberg.

Kulicke, M./Wupperfeld, U. in collaboration with Hemer, J./Traxel, H./von Wichert-Nick, D. (1996): Beteiligungskapital für junge Technologieunternehmen (BJTU). Ergebnisse eines Modellversuchs. Heidelberg.

Kurz, R./Röger, W./Zarth, M. (1990): Existenzgründungshilfen von Bund und Ländern. Eine Wirkungsanalyse der Programme im Hinblick auf Wettbewerb, Produktivitätswachstum und Beschäftigung. Gutachten im Auftrag des Bundesministers für Wirtschaft. Tübingen.

Picot, A./Laub, U.-D./Schneider, D. (1989): Innovative Unternehmensgründungen. Eine ökonomisch-empirische Analyse. Berlin.

Pleschak, F./Rangnow, R. (1995): Ergebnisse des BMBF-Modellversuchs "Technologieorientierte Unternehmensgründungen in den neuen Bundesländern" der Jahre 1990 bis 1994. Working paper. ISI: Karlsruhe, Freiberg.

Schmude, J. (1994): Geförderte Unternehmensgründungen in Baden-Württemberg. Eine Analyse der regionalen Unterschiede des Existenzgründungsgeschehens am Beispiel des Eigenkapitalhilfe-Programms (1979 bis 1989). Stuttgart.

Von Freyend, J.-E./Eberstein, H.-H./Kreklau, C. (Eds.) (various years): BDI Handbuch der Forschungs- und Innovationsförderung. Loose-leaf publication with continuous additions. Köln.

Von Wichert-Nick, D. with the collaboration of Kulicke, M. (1994): Ökonomische Entwicklung und Unternehmensstrategien junger Technologieunternehmen. Ergebnisse einer Befragung von im Modellversuch "Beteiligungskapital für junge Technologieunternehmen" (BJTU) begünstigten Unternehmen. Working paper. ISI: Karlsruhe.

Weiss, M. (1992): Finanzierungsfragen. In: Hölters, W. (Ed.): Handbuch des Unternehmens- und Beteiligungskaufs. 3rd fully revised and enlarged edition. Köln, 199-257.

Wupperfeld, U. (1994): Strategien und Management von Beteiligungsgesellschaften im deutschen Seed-Capital-Markt - Ergebnisse einer empirischen Untersuchung von 33 Beteiligungsgesellschaften und Banken. Working paper. ISI: Karlsruhe.

Wupperfeld, U. (1996): Management und Rahmenbedingungen von Beteiligungsgesellschaften auf dem deutschen Seed-Capital-Markt. Frankfurt am Main.

Zeitschrift für das Gesamte Kreditwesen (Publ.) (1993): Die Finanzierungshilfen des Bundes, der Länder und der internationalen Institutionen. Gewerbliche Wirtschaft. Frankfurt am Main.

The Venture Capital Market in Germany

Udo Wupperfeld

1.   Development of the German Venture Capital Market

The German business investment or venture capital market came into being in the 1960s when, against the background of increasing discussion on the equity deficit and the falling ratio of equity to total assets of small and medium industry, banks set up the first **business investment companies**[1]. These had the aim of making equity or equity-type funding available to small and medium-sized enterprises which were not quotable on the Stock Exchange. However, contrary to the expectations placed in them, they did not invest in small or young enterprises.

A second phase of the German venture capital market began with a wave of foundations in the early 1970s. Since 1970, the Federal Ministry of Economics (BMWi), as administrator of the European Recovery Programme (ERP) Special Fund has offered low-interest re-financing and guarantees for the commitments of non-profit guarantee organisations to small and medium-sized firms. The intention was, by extending the ERP programmes, to stimulate the founding of business investment companies with an element of (regional) economic promotion, thus helping to secure existing small and medium enterprises in their region. When it was found that the private companies were making less use than had been expected of the ERP special fund offers, the Laender (federal states) spurred the initiation of the **Mittelständische Beteiligungsgesellschaften (MBGs)**, which commenced their business activities in the 1970s and the 1980s in all the federal states. The most active of these were - and still are - the MBGs in Baden-Württemberg, Hesse and Bavaria, whereas the MBGs in the other Laender already reduced their investment activities in the second half of the 1970s, or in some cases ceased them completely (cf. Mayer/Müller 1991).

---

1   The Deutsche Beteiligungsgesellschaft m.b.H. (DBG) in 1965, the Allgemeine Kapitalunion GmbH & Co. KG (AKU) in 1966, the KBG Kapitalbeteiligungsgesellschaft m.b.H. in 1968, the Gesellschaft für Beteiligungen und Kapitalverwaltung m.b.H. & Co. (GeBeKa) in 1969, the Beteiligungsgesellschaft für die deutsche Wirtschaft m.b.H. in 1969.

However at that time - due to the ERP guidelines - the MBGs did not invest in new technology-based firms (NTBFs) either, so that there was no supply of investment capital for these firms. In particular, there was a lack of venture capital which was oriented towards long-term capital gain and which, therefore, did not put a strain on the liquidity of a portfolio enterprise the first few years of its development.

The German business investment capital market entered its third phase in the early 1980s. This was precipitated by information and discussions on the successful American venture capital model, in which equity plus management support were offered to technology-based enterprises which were young and had growth potential. In Germany, as in other countries, the success of companies like Apple and Genentech, and the figures on impacts on the turnover, profit and workforce of enterprises financed by venture capital, generated a **"venture capital euphoria"**. Finally, 1983 witnessed the birth of the German "venture capital scene", as companies styled on the US model, such as International Venture Capital Partners S.A. Holding, Luxembourg (IVCP) and Techno Venture Management, Munich (TVM), were set up by banks and industrial enterprises. At the beginning of the 1980s, the classic German business investment companies also participated increasingly in the early development phases of technology-based firms. In the second half of the 1980s, foreign venture capital companies also entered the German market.

With this third phase, a comparatively rapid **development of the German business investment market or venture capital market** began. Starting from 1965, it took 20 years for the national portfolio to reach one billion DM; four years later this was two billion DM, and in 1995 it topped 5.6 billion DM. From 1983 to 1995 the volume of the market increased seven-fold, with the number of participations increasing 2.8 times to just under 3,000. The growth rates have been particularly high till the beginning of the 1990s with a slow down in the last years (see Table 19) (cf. BVK 1994, 1995).

Table 19: **Long-term development of the German business investment capital market** (BVK members only)

Year	1983	1984	1985	1986	1987	1988	1989
*)	785	867	1,156	1,360	1,592	1,974	2,577
**)	1,069	1,138	1,288	1,429	1,583	1,536	1,752

Year	1990	1991	1992	1993	1994	1995
*)	3,221	4,029	4,553	4,982	5,372	5,632
**)	1,977	2,298	2,455	2,657	2,789	2,990

*)   Volume of participations (in million DM)
**)   Number of participations

Source: BVK 1994, 1995

However, both in the portfolio of the classical business investment companies and in the new venture capital companies, many investments in NTBFs either failed or did not develop according to plan. This was mainly a consequence of the **limited experience of the investment managers in selecting and consulting NTBFs** (cf. Schmidt 1988). The initial euphoria soon gave place to disillusion. Some of the venture capital companies founded in the early 1980s did not even really get off the ground; others ceased their business activities after a few years (cf. Bräunling et al. 1989).

The high failure rate, and the relatively high expenses of due diligence and management support for NTBFs relative to the size of the investment, fairly soon led to a **re-orientation of investment policy**. The German business investment companies concentrated on expansion financing, commitments to established enterprises in traditional sectors and management buy-outs. Many companies cut down their commitments to NTBFs or ceased them altogether. Thus by 1989, there was no longer any significant supply of risk-bearing capital for these enterprises. Only the MBGs in Baden-Württemberg and Hesse were financing NTBFs, within the framework provided by the ERP re-financing programme or the re-financing offers of the relevant regional government.

As well as the change in portfolio structure, the business investment companies were increasingly aiming to **derive a regular income from their investments**, in order to cover their management costs. The change in target groups and in the size of investments was also accompanied by a change in support strategies, i.e. the scope and intensity of portfolio management support diminished ("hands off" rather than "hands on"). Despite existing differences and investment focuses, this led *de facto* to a similarity in the investment policies of German business investment companies which levelled out the differences between venture capital companies and other forms of investment companies (cf. Frommann 1992).

For a few years now, however, there has again been a "movement" in the early stage segment of the German **venture capital market**. A significant role in this was played by the pilot scheme "Business investment capital for new technology-based firms " (BJTU), which created favourable conditions for the founding of several seed capital companies and has also contributed to encouraging a number of existing business investment and venture capital companies to participate once more in the early development phases of NTBFs (cf. Harnischfeger et al. 1992). Nevertheless, seed and start-up financings still only occupy a place of minor importance in the total portfolio today, with a share of approximately eight percent (in terms of investment volume) in total portfolio and 14 percent in gross investment in 1995 (cf. Table 20). By comparison, the first share was still about 25 percent in 1985 (Frommann 1993). By far the greatest volume of the total German portfolio is invested in expansion financing, with management buy-outs and management buy-ins ranking second, quite a long way behind. Seed and start-up investments constitute a substantially higher proportion when considered in terms of the total number of investments, with a share of approximately 31 percent (total portfolio) resp. 38 percent (gross investment in 1995).

Table 20:   **The German venture capital market in 1995, by financing phases**

Financing phase	Total portfolio				Gross investment			
	million. DM	in %	number	in %	million DM	in %	number	in %
Seed	102.90	1.83	204	6.82	30.66	2.90	49	7.38
Start-Up	356.34	6.33	738	24.68	114.35	10.80	203	30.57
Expansion	3,666.17	65.10	1,681	56.22	568.74	53.72	290	43.67
Bridge	290.10	5.15	18	0.60	49.70	4.69	18	0.60
MBO/MBI	992.67	17.63	240	8.03	214.74	20.28	58	8.73
Turn-Around	148.50	2.63	44	1.47	62.92	5.94	11	1.66
(Altogether)	5,556.77	98.67	2,925	97.82	1,041.11	98.33	617	92.92
Total (incl. no details)	5,631.76	100.00	2,990	100.00	1,058.76	100.00	664	100.00

Source: BVK 1996

In Germany, the private and public banks in particular act as investors, providing considerably over half the volume of funding supplied by all business investment capital companies (cf. Table 21). Insurance companies and industrial enterprises, which usually have a sceptical attitude towards venture capital, the public sector and other investors play only a subordinate role. Although in a European comparison credit institutions also emerge as the most important investors of business invest-ment capital companies, they do not have for Europe as a whole the same central importance that they have in Germany. The second most important type of investors in Europe are the pension funds, which in Germany do not figure as investors at all (cf. evca 1993).

In the new federal states the venture capital market is just in the process of forma-tion, so that it is not yet possible to make statements about the main investors of the future there. On the one hand, West German and foreign investors have launched extensive funds; on the other hand, MBGs have been set up in all the new federal states. The first investments in the new federal states were made in 1990; however the level in 1995 (743 Million DM investment volume, 497 participations) was still low, both relative to the old federal states and relative to the demand for investment capital. NTBF were practically not financed at all. The causes for the low level of commitment of business investment companies so far are to be found in the numer-ous unresolved issues of ownership, in administrative barriers and in the generally very high risk factors associated with enterprises seeking capital (cf. evca 1992). Since the mid 1990s, however, the engagement in seed financing and the provision of business investment capital, increased (e.g. by seed capital companies with a regional focus like the Phoenix Venture Fund in Saxony).

Table 21:   Sources of capital, by sectors, 1995

Investor	Volume (in million DM)	%
Banks	5,275.89	57.21
Insurance companies	728.02	7.89
Pension funds	789.90	8.57
Industry	907.20	9.84
Private	420.10	4.56
Public	728.12	7.90
Others	234.48	2.53
(Altogether)	9,083.71	98.50
Total (incl. no details)	9,221.87	100.00

Source: BVK 1996

2.   Types of German Business Investment Companies

Today there are various types of investment companies on the German venture market (cf. Table 22), which can be classified according to the following features:

- their initiators and investors,
- their business aim in investing (high returns, promoting the economy),
- the emphasis of their investment with regard to financing phases,
- the type and extent of management support they offer, and
- the financing instruments they prefer (e.g. direct or silent participation).

Table 22:   Types of German business investment companies

Features	Type of investment company	Venture capital companies of banks, insurance companies	Venture capital companies of savings banks	MBGs	"classic" venture capital companies	Seed capital companies
Initiators/investors		banks, insurance companies	savings banks, people's banks and rural cooperative banks and their umbrella organisations	"Laender" (federal state) governments, other public bodies, credit institutions	credit institutions, industry, investment managers	investment managers, industry, credit institutions
Purpose of investment		profit-oriented (current income)	profit-oriented (current income), to some extent also promotion of the economy	promotion of the economy	profit-oriented (capital gain)	profit-oriented (capital gain)
Investment emphasis		growth enterprises	growth enterprises	small and medium sized enterprises	(innovative) growth enterprises	seed and start-up phase of NTBFs
Importance of NTBF		none	small	small	small	dominant
Support services		few	few	very few	extensive	extensive
Form of participation		no preferred form	silent, sometimes direct	silent	direct	direct and silent
Main foundation period		from 1960s, mainly in 1980s	from 1960s, mainly in 1980s	1970s and 1980s 1992 onwards in the new federal states	early to mid-1980s	from end of 1980s
Geographic orientation		national	administrative area of savings bank or rural cooperative bank or umbrella organisation	relevant federal state	national, to some extent international	part regional, part national, a few also international
Desired form of exit		repayment of silent partnership by enterprise, resale to third party, Stock Exchange	repayment of silent partnership by enterprise, resale to industry, Stock Exchange	repayment of silent partnership by enterprise	Stock Exchange, resale to industry	resale to VC companies, industry, Stock Exchange

Source: own compilation

Investment companies of banks and insurance companies

These are business investment companies (ICs) with a dependent status, founded by banks or insurance companies. The **ICs of banks** pursue yield-related goals, concentrating mainly on current income from their investments. To some extent they also pursue strategic aims such as the initiation and intensification of customer relations (portfolio enterprises as potential borrowers). Since the early 1980s, aspects such as gathering experience in the evaluation of innovative business concepts for credit financing, and improving the bank's image, have played a role (cf. Büschgen 1985; Grisebach 1989). The venture capital companies of banks concentrate their investments in the financing of established growth enterprises. They practically never invest in small or new enterprises, whose demands in terms of investment volume are small. The reasons for this are the fact that risks are difficult to assess, and the unfavourable ratio of the high costs involved in selection and support, to the small size of the investments. The ICs of banks provide only limited management support services for their portfolio enterprises.

Within the limits set by their general aims and by the Law on the Supervision of Insurance Companies (VAG, Versicherungsaufsichtsgesetz, particularly § 54), insurance companies have also participated in setting up ICs or have founded their own daughter companies. Due to the low synergy effects between business investment and insurance activities, **the ICs of insurance companies** aim exclusively for the highest possible returns. In other respects their participation policy is very similar to that of the ICs of banks (cf. Kulicke et al. 1993).

Business investment companies of the savings banks

ICs of the German savings banks, people's banks, rural cooperative banks and their umbrella organisations (mainly regional banks and central giro institutions) also pursue aims which are primarily acquisitive, but in addition they sometimes have aims relating to the promotion of the regional economy. In the latter case, they mostly offer silent partnerships under favourable conditions to SMEs located in their region. For this they often make use of the re-financing offer of the ERP Fund. At the beginning of the 1980s, the ICs of the savings banks demonstrated a high degree of commitment to the financing of NTBFs in connection with the setting-up of technology and incubator centres which took place at that time. Today, however, like the ICs of the banks and insurance companies, they are mainly investing in growth enterprises; only a few regional companies participate in NTBFs to any notable extent.

"Mittelständische Beteiligungsgesellschaften" (MBGs)

The **Mittelständische Beteiligungsgesellschaften/Laender funds (MBGs)** are business investment companies with an economic policy mandate, which engage in silent partnerships in SMEs. MBGs - as "self-help organisations for industry" - are carried mainly by chambers of industry and commerce, regional credit institutions, and work in close regional co-operation with credit institutions. They do not have any own funds to speak of, but re-finance themselves primarily from promotion programmes of the Federal Government and the Laender, of which the ERP funds are most important for the MBGs. Thus their conditions of investment (with regard to the amount invested, age of firm, form of participation, regulation and time schedule of repayments) are determined largely by the procedures of the ERP programme. In case of failure of supported firms, the MBGs can fall back on guarantees of the Federal Government and the federal state concerned, via guarantee banks or guarantee associations (Schütt 1993).

Thus most MBGs concentrate their investments in the financing of SMEs. The MBGs in Baden-Württemberg, Hesse and Bavaria, and the Innovation Fund of Berlin, also finance the early development phases of NTBFs, making use of the BTU programme and the relevant programmes of the Laender governments.

"Classic" venture capital companies

Unlike the investment companies of banks, savings banks and insurance companies, **venture capital companies** are independent suppliers of venture capital whose investors are recruited mainly from industrial firms, credit institutions and investment managers. In the last few years, venture capital companies from abroad have increasingly been setting up daughter companies in Germany. Whereas in the early 1980s venture capital companies primarily financed NTBFs, today they only invest in this group of firms to a small extent. They concentrate mainly on financing the expansion phases of small and medium sized enterprises (including technology-based enterprises), management buy-outs and turn-arounds, since in these investments the relationship between the investment volume and the expense incurred by evaluation and support of the portfolio enterprises is more favourable than with NTBFs.

Venture capital companies are exclusively profit-oriented and mainly engage in direct participations, expecting to realise their profits less in the form of regular dividends than as capital gain on resale of the investment. They advertise themselves as being in a position to provide their portfolio enterprises with comprehensive management support, due to their international networks and interdisciplinary management teams. Since this type of investment companies first became active on the German market in the 1980s, there is as yet no long-standing tradition of direct, capital gains-oriented investments in Germany.

Seed capital companies

Seed capital companies are a special form of venture capital companies. They differ from them in the emphasis of their investments, which are concentrated on the early development phases of technology-based firms. These relatively young companies are mostly independent and can be traced back mainly to two types of initiators: on the one hand, these are managers, usually with a "foundation-oriented background" who have been stimulated by public promotion programmes for seed capital commitments (such as the BJTU pilot scheme or the European Pilot Scheme for the Stimulation of Seed Capital) to set up a new investment company. On the other hand, seed capital companies sometimes result from the initiatives of credit institutions (cf. Kulicke et al. 1993).

Seed capital companies aim for a high return on their investment, which they expect to realise in the form of capital gain. However, at the same time some of them pursue the aim of promoting the economy. Seed funds finance their portfolio enterprises mainly in the form of direct participation, but sometimes as silent partners, and offer them active, extensive management support. This investment policy means that, of all the types of investment companies, seed capital companies correspond most closely to the "classical" venture capital concept. Since NTBFs have great difficulties in acquiring investors in their early stages of development, seed capital companies fulfil the function of "knock-on financing". In this case seed funds offer "intelligent" equity capital, i.e. combined with intensive support.

3. Frame Conditions and Barriers to the Development of the German Seed Capital Market

Although the German venture capital market has shown evidence of rapid growth over the last few years, the segment of early phase financing of NTBFs is still relatively underdeveloped. In the following section, the **barriers to development of the German seed capital market** are discussed. This includes a review of the taxation system, the legal requirements relating to investments, the German Stock Exchange, alternative financing options for enterprises, and socio-economic factors, in an illustrative comparison with the USA.

3.1 Taxation

Taxation is centrally important to all financing and investment decisions, since it influences an enterprise's choice of financing medium and affects the attractiveness of investment options. Thus investigations in the USA, Great Britain and Canada confirm a negative correlation between the size of capital gains tax and investments in venture capital companies (cf. for instance Poterba 1989; McMurtry 1986; Eisenhardt/Forbes 1984). In a fiscal evaluation of venture capital financing, three actors have to be considered: the investor, the investment company and the portfolio enterprise. The monetary flows between these three entities are shown in Figure 16.

The **fiscal treatment of the investor's stake in the investment company** depends on the form of the investment. If this is a participation in an incorporated firm, it is subject to company tax. If it is in the form of a participation in a general commercial partnership (offene Handelsgesellschaft, OHG), a limited commercial partnership (Kommanditgesellschaft, KG) or an atypical "silent" investment, then there is, from the fiscal point of view, a co-entrepreneurship, which is not subject to company tax.

The German fiscal system does not provide for any tax benefits for investments in business investment companies. On the side of the investment company, the acquired capital is subject to capital tax.

Figure 16: **Monetary flows in investment financing**

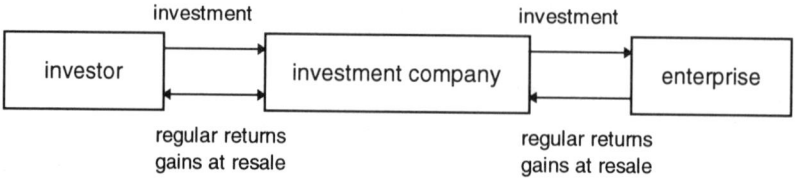

Regarding the **stake of the business investment company in the portfolio enterprise,** the same regulations apply as for the investor's stake in the investment company. Since from the standpoint of the investment company the investment is purely a matter of asset redeployment, it does not qualify for tax benefits either. For the portfolio enterprise, the sale of shares and the conclusion of a silent partnership are not relevant from the viewpoint of taxation. However, for the enterprise there is a disadvantage compared with the acquisition of long-term dept capital, in the sense that when working out the taxable base for the trade tax, 50 percent of the permanent debts from the trading capital, and 50 percent of the interest from the trade earnings, have to be added in. This means that equity capital and profits are more harshly treated by trade tax than debt and interest on borrowings.

The **taxation of regular returns from participation in a business investment company** also depends on the form of investment. In the case of an incorporated firm, the investor's positive or negative income (from capital assets) is subject to the income tax regulations that are relevant for him, with corporation income tax and the retained capital gains tax being set off. These are paid by the investment company since, as an incorporated firm, it is an independent taxable entity. If there is a co-entrepreneur type of participation in an investment company, taxes are incurred by the investor, and in this case he can ask for the profits to be taxed as income from business establishment and can claim for losses.

Fiscally speaking, the **resale of the investor's share in an investment company** is handled as follows: profits and losses of a natural person from the resale of a participation in an incorporated firm are of no importance from the viewpoint of taxation. If the share is in excess of 25 percent, the reduced taxation rate according to § 34 of the German Income Tax Law comes into force. In the case of a co-entrepreneur participation, the gain on disposal is subject to income tax; natural persons only have to pay half the average rate. The resale gain of legal entities is subject to full taxation, without regard to the form or size of the participation.

The regular returns of the investment company from its participation in a portfolio enterprise, and the resale of shares, are subject to the same tax regulations as the regular returns and the gain (or loss) on disposal resulting from the participation of legal entities in an investment company. From the viewpoint of the portfolio enterprise, the regular distribution of dividends (as a regular use of income) is neutral with regard to taxation, as is the resale of the investment company's shares.

In summary, it can be stated that **German fiscal law does not envisage any special regulations or tax benefits for the financing of enterprises by business investment companies** (cf. Wupperfeld 1993). Nevertheless, tax advantages can be gained by making use of the fiscal options available, i.e. by selecting a legal form for the transaction which is suited to the ideal course of the investment financing. If an investment company is still at an early stage in its development and incurs losses from its investments, a co-entrepreneurship option (for instance "GmbH & Co. KG"[2]) offers investors the possibility of tax-free allocation of losses. If the investment company achieves the net income area, it is advantageous for it to change its status to a public limited company, because in this way the fiscal disadvantages of capital gain taxation at divestiture can be avoided (cf. Fischer 1987).

The question of the extent to which the present German tax legislation hinders the development of the seed capital market in comparison to the USA cannot be exhaustively discussed here. There are (a few) surveys on this subject (for example, Nevermann/Falk 1986; Swoboda/Zecher 1985; Weichert 1987) which indicate that, in comparison with the USA, the German venture capital market is fiscally disadvantaged; however, these have some basic deficits, since

- they only consider certain aspects of the existing US American and German tax systems (e.g. possibilities for writing-off assets, in order to reduce taxable gains),

- they do not include the fiscal treatment of alternative investment options for investors in their analyses,

---

2 GmbH & Co. KG: a limited commercial partnership (KG) formed with a limited liability company (GmbH) as general partner and the members of the GmbH, their families, or outsiders, as limited partners. In accordance with § 19 of the German Commercial Code it is formed as a family-name firm.

- they are based, with regard to the USA, on the combined legal form of "SBIC/Subchapter-S-Corporation" which, although eligible for particular tax benefits, has a limited importance in practice.

3.2 The Legal Requirements Relating to Institutional Investors

The legal conditions applicable to the investments of institutional investors determine the extent to which certain groups of investors are allowed to put the funds entrusted to them by third parties into business investment companies, and thus influence the flow of capital into investment companies. The following section therefore investigates the possibilities open to pension funds, insurance companies, public investment funds and banks to invest in such funds.

There is a consensus in the literature that the sudden upswing of the venture capital industry in the USA in the 1980s was attributable mainly to the activities of the **pension funds** (cf. Bygrave/Timmons 1992). In Germany, on the other hand, there are practically no pension funds, since the money for employee pension schemes usually remains within the same firm as company pension reserves, due to the associated tax advantages.

Since 1987 another potential group of investors, the **insurance companies**, have been able to invest five percent of the cover funds and 6.25 percent of the remaining restricted assets in the form of business investment capital. However, so far this extension of investment possibilities has only led to a slight increase in the commitment of insurance companies to such funds.

A structuring of business investment capital investment companies as **public investment funds** would have the advantage of higher fungibility. This structure is not possible in Germany, however, as the investment options of funds of this kind are limited to listable enterprises. The Act for improving the framework conditions for institutional investors which came into force on January 1, 1987 was a legislative attempt to make it easier for investment trusts and insurance companies to acquire equity investments. However, the impacts on the seed capital market have been negligible so far (cf. Fetzer 1990).

Thus in Germany **banks** represent the most important group of investors. This constitutes a considerable difference from the situation in the USA where, due to the system of functional separation in the financial services sector, commercial banks are not allowed to participate in activities outside the banking sector, a bar which includes venture capital business. The commitment of German banks is limited, however, by their low ratios of equity to total assets, which do not permit any sizeable losses, e.g. from high risk investments. Moreover venture capital investments together with other long-term investments, is not allowed to exceed the banks' own funds. This explains the careful, risk-rejecting policy of banks, designated as „Prudent Banking" (cf. Büschgen 1985; Stedler 1993). Nevertheless, German banks emerge in an international comparison as relatively willing to take risks (cf. Quillmann 1987).

All in all it can be stated that, irrespective of these limitations, the possibilities of institutional investors for investing in business investment funds in Germany are by no means exhausted. Therefore state regulations probably do not have any substantial negative impact on the development of the German venture capital and seed capital market. In other words, institutional investors could theoretically invest far more in business investment companies than they actually do. Nevertheless, investment regulations which were generally more liberal could lead in the long term to an innovative, diversified finance sector, and could prepare the way for a more rapid acceptance of seed capital by the finance sector (cf. Wupperfeld 1994).

3.3   The German Stock Exchange

The structure of the German Stock Exchange - and particularly the listing regulations for individual stock market tiers - exert a direct influence on how, and at what price, shares can be sold, and thus directly influence the profit chances of business investment companies. Thus empirical surveys in the **USA** confirm a very strong correlation between the development of the OTC market ("Over The Counter", the stock exchange tier that represents investment companies' most important exit route in the USA) and the returns achieved by investment companies. These studies also confirm that without the stock exchange exit option, venture capital would not be a viable business proposition in the USA (cf. Bygrave/Timmons 1992).

In the **Federal Republic of Germany,** trading in securities takes place on the official market, on the regulated market (a second, intermediate tier that has formed in accordance with European stock exchange guidelines 1987) and in the off-board trading. The latter is unofficial and exists on a purely private-law basis. However, it only meets with a low degree of acceptance due, among other factors, to open questions with regard to legal aspects of liability. The official market is intended mainly for large, established enterprises and for bonds of the Federal Government and other issuers.

The regulated market, initiated in 1987, is definitely suitable for the disinvestiture of investment companies' commitments, because of its low listing requirements with regard to the age of the firm, the minimum total volume of issues, the distribution of shares and the listing charges (cf. Stedler 1987). However, until now this form of exit has not played any substantial role for German venture capital companies. Fetzer ascribes this mainly to the German universal bank system and the competitive power of the banks (cf. Fetzer 1990). The universal bank system permits credit institutions to handle issues (cf. Weichert 1987), so that unlike the USA new issues are almost exclusively handled by banks. Since with small enterprises the brokerage fees are relatively small and the tightness of the market makes price management more difficult, the attractiveness of an issue goes down with the size of the firm. Moreover, with new and small enterprises the liability relating to statements made in the issuing prospectus represents a high risk for the issuing bank. Credit institutions thus prefer to finance small firms via credits, all the more so because there is practically no competition from the Stock Exchange in the area of equity provision. Banks will only make efforts to support in the initial public offering of a firm from a certain size upwards. This business strategy is concealed behind the stringent minimum requirements for potential issuing firms (cf. Ingram/Miles 1984; Hausberger 1984).

Since the organised capital market in Germany has not so far played any part as an "exit route" for participations in NTBFs, there is no "knock-on" effect which would also favour the development of an upstream capital market segment, as the venture capital market. In summary, it can be said that the **non-existence of a Stock Exchange exit route** diminishes investment companies' chances of profits from participations in NTBFs. It thus constitutes a frame condition which **impedes the development of the seed capital market**. Because of this, German venture capital companies are planning initial public offerings of NTBFs on the US American computer stock exchange NASDAQ (for instance, TVM, Techno Venture Management GmbH & Co.; cf. Ludsteck 1994).

3.4 Alternative Financing Options Open to New Technology-Based Firms

It can be assumed that the demand of NTBFs for venture capital is dependent on the financing alternatives available to them. Therefore in the following section the "competing products" of long-term bank loans, promotion funds and informal capital, and their impacts on the financing behaviour of NTBFs, are discussed.

The sources of finance primarily available in Germany to SMEs trying to raise capital are several very liquid sources in the form of loan capital, the main bulk of these being **long-term bank credits**. In Germany, the banking sector plays a dominant role in the financing of enterprises (cf. Lucas 1994), so that in comparison with the USA, business investment companies in Germany have to face much stronger competition from banks. German credit institutions are more willing to take risks than those in the United States. For instance, requirements regarding the amount of equity capital are much lower when giving credits here in Germany than in the USA. This can be regarded as a consequence of the high savings of private households and the networking of the finance markets, ensuring a high liquidity of the banks and a correspondingly high readiness to give credits (cf. Schramm 1988). Banks have also started to enlarge their service spectrum, for instance by advisory services (cf. Rüschen 1990) or special innovation financing departments.

This results in a smaller demand for business investment capital than in the USA. In this context, Fetzer even speaks of an over-supply of loan capital which decreases the demand for equity capital (cf. Fetzer 1990). However, Fetzer's investigation is not explicitly confined to the financing of the early development phases of NTBFs, but relates to SMEs in general. For the latter categories, the acquisition of loan capital does not present so many problems, whereas banks are cautious about giving loans to very young, high-risk technology-based firms. Thus it is certainly not possible to speak of an over-supply of loan capital in the case of NTBFs. Nevertheless, one can go along with Fetzer's statement that the dominant role of the banking sector in the financing of enterprises has generally been an impediment to the development of the venture capital market and to the formation of a venture capital culture.

In the old federal states there was until a few years ago a strong promotion offer for small and medium and new firms compared with the USA. Thus in 1988 alone, the funds used by the Reconstruction Loan Corporation (KfW) and the German Equalization Bank (DtA) in the **promotion of small and medium industry** totalled more than eight billion DM (cf. Müller-Kästner 1989). Considerably more than 300 NTBFs were supported in the 1980s under the pilot scheme "Promotion of New Technology-Based Firms" (TOU) (cf. Kulicke u.a. 1993). There is still an extensive promotion offer for NTBFs in the new federal states, including principally the former pilot scheme "Promotion for New Technology-Based Firms in the New Laender" (TOU-NBL), which supports NTBFs by extensive subsidies (cf. Bachelier 1993), as well as the follow-on measure FUTOUR (cf. the contribution on the promotion of NTBFs in this reader).

The impacts of this promotion funding on the demand for venture capital have to be considered individually. It may certainly be supposed that the overall effect of the promotion offer for new, small and medium enterprises on the development of the venture capital market has been inhibitive. However, this does not explain the small importance of the seed segment within the venture capital market as a whole. In the case of NTBFs, promotion may actually have had the opposite effect, by helping to initiate innovative foundations and make these enterprises into interesting objects for participation. In other words, the promotion fulfils the function of "knock-on financing": technology-based firms are put in a position where they can carry out their development tasks and finance the build-up of their technological basis and their first marketing activities. Once they have reached this development phase, NTBFs represent less of a risk for business investment companies, especially as the capital base of the enterprise is also broader.

An important source of financing for the seed phase of newly-founded firms in the USA is **informal investment capital**. This is provided by members of the family, friends or independent private persons (Business Angels). In Germany, unlike the USA, a "seed culture" of this kind is, as yet, practically non-existent (cf. Frommann 1993; on Business Angels in the USA and Great Britain, see e.g. Harrison/Mason

1991; Wetzel 1983; Wetzel 1987; Mason/Harrison 1992). Moreover, in Germany commitments of this kind are regarded more as capital investments and are not so much made with the intention of participating in management. Thus many NTBFs in Germany, whose founders do not have sufficient own funds or cannot acquire promotion funding, have to do without this form of knock-on financing, which may also include support for the firm in elaborating a business plan.

All in all, it can be stated that the impacts of financing alternatives on the development of the seed capital market cannot be clearly determined. On the one hand, the dominance of banks in the financing of enterprises, and the extensive promotion offer available to small and medium-sized firms, have certainly had an inhibiting effect on the development of the venture capital market as a whole. On the other hand, the promotion of NTBFs may be said to have had a stimulating effect.

3.5 Socio-Economic Factors

As well as fiscal, legal and economic factors, the development of the venture capital market is influenced by societal, social and psychological aspects. One important factor here is differences in mentality, i.e. national value judgements and the value judgements of enterprises and managers, which are more conducive to the development of venture capital in the USA than in Germany (cf. Fetzer 1990). Empirical surveys show that American attitudes differ fundamentally from European attitudes, and particularly from German ones. They confirm the picture of security-oriented, hierarchically-minded Germans with a strong consumer and leisure orientation. By contrast, in the USA willingness to take risks, and career ambitions are much more strongly developed - as one might expect from American history, permeated as it is by the pioneer spirit and the will to succeed (cf. Grimm 1985; Hierl 1984; Ludsteck 1993).

Although, in international comparison, differences in values and capabilities emerge as much smaller for the group of German entrepreneurs and managers than for the group of Germans as a whole (cf. Getas 1989; Grimm 1985; Mohler 1989), this only seems to apply up to a point to the founders of enterprises. Thus in the USA, a high income and the ambition to achieve were the main motives for founding an enterprise. The dominant considerations in the Federal Republic of Germany, on the other hand, were the wish for more freedom of decision and freedom of action, the wish to be independent and realize ones own ideas. German founders also mostly wanted to remain sole owners, even if this would limit the growth chances of their enterprise. This basic attitude stands in the way of a higher acceptance of venture capital, and in particular of direct, "hands-on" investments (cf. Ludsteck 1994).

Although the climate for setting-up enterprises in Germany improved substantially in the 1980s, resulting in an increase - also partly attributable to the favourable economic trend at that time - in the number of foundations, nevertheless the economic and social frame conditions in Germany do not favour the founding of firms so much as conditions in the USA. Thus the well-developed social "safety net" in Germany, and the relatively high salary levels in management, do not create such strong incentives for personal initiatives and independence (cf. Dubini 1988); as a consequence, the conditions for seed capital are less favourable.

Although German venture capital managers in the 1990s are much more highly-qualified than in the 1980s, a **lack of professional fund managers** is often still remarked on. This applies particularly to the management of participations in NTBFs, in which the assessment and support of the firms make particularly intensive demands on investment companies. Empirical surveys have shown that venture managers in the United States have a more varied background and a higher level of qualification than their German counterparts (cf. Fetzer 1990; Schröder 1992). The high degree of professionalisation in this sector, also expressed in a stronger internationalization (cf. Frommann 1993), is understandable in view of the much longer tradition of venture capital in the USA, which has even led to the creation of a career orientation known as "Venture Capital Investment Manager". Since the success of investment funds depends largely on the quality of their management (cf. Bygrave/Timmons 1992) and, moreover, experienced investment managers are able - as a survey in the USA has shown - to acquire more capital (cf. Klemm 1988), it can be supposed that these factors have also had an inhibiting effect in Germany on the development of the venture capital market in general, and on the development of the seed capital market in particular.

4.   Summary: Frame Conditions and Barriers to Development

In the past, a number of factors have inhibited the development of the German venture capital market. The main inhibitors have been:

- the German universal bank system and the dominant role of credit institutions in the financing of new, small and medium-sized firms; there are several alternative financing options open to NTBFs in Germany; NTBFs are not necessarily aware of the venture capital option, nor do they accept it to any great extent;

- the lack of possibilities for an exit via official quotation on the Stock Market. This particularly affects new and high-risk enterprises, and means that the conditions for venture capital companies to reap high returns from their participations in NTBFs are not fulfilled;

- a lack of tax advantages for venture capital;

- socio-economic factors, such as the wish for independence and founders' low acceptance of partners with a say in business decisions. In addition, Germans' relative reluctance to take risks, and a comprehensive social „safety net" linked to the employed status, inhibit the transfer of management talent to smaller firms. The financing instrument of venture capital is also largely unknown to many founders of technology-based firms. Business investment companies frequently complain of too little demand from investment objects with promising chances.

Thus it was only in the 1980s that any noteworthy market volume grew up; however, most investment companies concentrate on financing the growth processes of established enterprises, management buy-outs, etc. In Germany, therefore, there is no long-standing venture capital tradition comparable to the USA. Due to this lack of a venture capital tradition and the relative newness of the seed capital market in Germany, there is still a lack of qualified professionals today. Up to now only a few funds have accumulated broad experience with NTBFs. Efficient networks and suitable instruments for evaluating and servicing NTBFs - which are prerequisites for higher returns on investments - have thus only developed in a few specialised companies.

Nevertheless, the conditions for participating in NTBFs have become decidedly better over the last few years through the promotion offered by the Federal Government's BJTU pilot scheme and its successor, the programme BTU. Without such massive promotion support, however, the venture capital market for NTBFs in Germany would not be economically viable. In other words, investment companies would not, in all probability, achieve a sufficient return on their investment. For this to be possible the other frame conditions, particularly the possibility of initial public offerings need to be further improved (cf. Wupperfeld 1996).

5.   Bibliography

Bachelier, R. (1993): Die neuen Fördermöglichkeiten des BMFT zur Finanzierung des Produktionsaufbaus und der Markterschließung (Phase II des Modellversuchs TOU-NBL) In: Bräunling, G./Pleschak, F./Sabisch, H.: Finanzierung des Produktionsaufbaus und der Markterschließung geförderter junger Technologieunternehmen in den neuen Bundesländern. Konferenz und Workshop am 16. und 17. März 1993 in Leipzig. Conference paper. ISI: Karlsruhe, 4-5.

Bräunling, G./Gerybadze, A./Mayer, M. (1989): Ziele, Instrumente und Entwicklungsmöglichkeiten des Modellversuchs "Beteiligungskapital für junge Technologieunternehmen" (BJTU). Working paper. ISI: Karlsruhe.

Bundesverband deutscher Kapitalbeteiligungsgesellschaften (BVK) (Publ.) (1994): Jahrbuch 1994.

Bundesverband deutscher Kapitalbeteiligungsgesellschaften (BVK) (Publ.) (1996): Jahrbuch 1996.

Büschgen, H.E. (1985): Banken und Venture-Capital-Finanzierung. In: Die Bank, No. 6, 284-292.

Bygrave, W.D./Timmons, J.A. (1992): Venture Capital at the Crossroads. Boston, Mass.

Dubini, P. (1988): The Influence of Motivations and Environment on Business Start-Ups: Some Hints for Public Policies. In: Journal of Business Venturing (4), 11-26.

Eisenhardt, K.M./Forbes, N. (1984): Technical Entrepreneurship: An International Perspective. In: Columbia Journal of World Business (19), 31-38.

Europe's Venture Capital Association (evca) (1992): Venture Capital in Europe. 1991 evca Yearbook. London.

Europe's Venture Capital Association (evca) (1995): Venture Capital in Europe. 1994 evca Yearbook. London.

Fetzer, R. (1990): Analyse internationaler Unterschiede im Volumen und in der Struktur von Venture-Capital-Aufkommen und -Anlage. Analysen zur Strategie ausgewählter Akteure im Netzwerk der jungen Technologieunternehmen. Berlin.

Fischer, L. (1987): Problemfelder und Perspektiven der Finanzierung durch Venture Capital in der Bundesrepublik Deutschland. In: Die Betriebswirtschaft (47), 8-32.

Frommann, H. (1992): Venture Capital in Deutschland - Rückblick auf ein Vierteljahrhundert. In: Bundesverband deutscher Kapitalbeteiligungsgesellschaften (BVK): Geschäftsbericht 1992, 101-106.

Frommann, H. (1993): Entwicklungstrends am deutschen Beteiligungsmarkt. In: Bundesverband deutscher Kapitalbeteiligungsgesellschaften (BVK): Jahrbuch 1993, 11-31.

Getas (1989): Die Deutschen als Europäer - Teilergebnisse aus ACE Anticipating Change of Europe und Getas-Report. Hamburg.

Grimm, E. (1985): Wertewandel-Konsumwandel. In: Planung und Analyse (12), 392-396.

Grisebach, R. (1989): Innovationsfinanzierung durch Venture Capital - Eine juristische und ökonomische Analyse. München.

Harnischfeger, M./Kulicke, M./Wupperfeld, U. (1992): Zum Stand des Modellversuchs "Beteiligungskapital für junge Technologieunternehmen" (BJTU) - Zwischenbericht zum 31.12.1991. Working paper. ISI: Karlsruhe.

Harrison, R.T./Mason, C.M. (1991): Informal investment networks: a case study from the United Kingdom. In: Entrepreneurship & Regional Development (3), 269-279.

Hausberger, H. (1984): Wiederbelebung der Aktie. In: Wirtschaftsdienst 1984, No. 7, 335-340.

Hierl, W. (1984): Venture Capital - auf deutsche Verhältnisse übertragbar? In: Kreditpraxis, No. 3, 5-8.

Ingram, D.H.A./Miles, D.K. (1984): Unternehmensfinanzierung in Großbritannien und in der Bundesrepublik Deutschland. In: Monatsberichte der Deutschen Bundesbank, No. 11, 35-46.

Klemm, H.A. (1988): Die Finanzierung und Betreuung von Innovationsvorhaben durch Venture Capital Gesellschaften. Möglichkeiten und Grenzen der Übertragung des amerikanischen Venture Capital Konzepts auf die Bundesrepublik Deutschland. Frankfurt am Main

Kulicke, M. et al. (1993): Chancen und Risiken junger Technologieunternehmen - Ergebnisse des Modellversuchs "Förderung technologieorientierter Unternehmensgründungen" (TOU). Heidelberg.

Kulicke, M. (1995): Hintergrundinformationen zur Pressekonferenz von Bundesminister Dr. Rüttgers anläßlich der Vorstellung des neuen Förderprogramms "Beteiligungskapital für kleine Technologieunternehmen" (BTU). Working paper. ISI: Karlsruhe.

McMurtry, B.J. (1986): Tax Policy Influence on Venture Capital. In: Landau, R./Jorgenson, D. (Eds.): Technology and Economic Policy. Cambridge, 137-151.

Mason, C./Harrison, R. (1992): The Financing of Technology-Based New Firms in the UK: The Role of Informal Venture Capital. In: Fraunhofer-Institut für Systemtechnik und Innovationsforschung (ISI)/Warwick Business School (Publ.): Proceedings of the Anglo-German Seed-Capital Workshop. Karlsruhe, Warwick, 58-86.

Mayer, M./Müller, R. (1991): Die Deutsche Wagnisfinanzierungsgesellschaft mbH (WFG) - Erfahrungen und Ergebnisse eines Modellvorhabens. Working paper. ISI: Karlsruhe.

Mohler, P.P. (1989): Der Deutschen Stolz: Das Grundgesetz. In: Informationsdienst Soziale Indikatoren, No. 2, 1-4.

Müller-Kästner, B. (1989): Das Mittelstandsprogramm soll größenbedingte Nachteile bei der langfristigen Fremdfinanzierung ausgleichen. In: Handelsblatt, 8.4.1989, 34.

Nevermann, H.;/Falk, D. (1986): Venture Capital. Ein betriebswirtschaftlicher und steuerlicher Vergleich zwischen den USA und der Bundesrepublik Deutschland. Baden-Baden.

Poterba, J.M. (1989): Venture Capital And Capital Gains Taxation. In: Lawrence, H. (Ed.): Tax Policy and the Economy, Vol. 3. Cambridge, 47-67.

Quillmann, W. (1987): Venture Capital in den USA und Deutschland. In: Die Bank, Nr. 12, 669-673.

Rüschen, T. (1990): Consulting-Banking: Hausbanken als Unternehmensberater. Wiesbaden.

Schramm, B. (1988): Finanzierung nicht emissionsfähiger mittelständischer Unternehmen. In: Christians, F.W. (Ed.): Finanzierungshandbuch, 2nd Edition. Wiesbaden, 563-576.

Schütt, F.H. (1993): Deutsche Bürgschaftsbanken in der Bewährung. In: Sparkasse, (110), 465-467.

Schmidt, R.H. (1988): Venture Capital in Deutschland - Ein Problem der Qualität? In: Die Bank, No. 4, 184-186.

Schröder, C. (1992): Strategien und Management von Beteiligungsgesellschaften: ein Einblick in Organisationsstrukturen und Entscheidungsprozesse von institutionellen Eigenkapitalinvestoren. Baden-Baden.

Stedler, H.R. (1993): Beteiligungskapital im bankbetrieblichen Leistungsangebot. In: Die Bank, Nr. 6, 347-351.

Swoboda, P./Zecher, J. (1985): Unternehmensberatung und Risikokapitalbildung. In: Betriebswirtschaftliche Forschung und Praxis (37), 402-420.

Weichert, R. (1987): Probleme des Risikokapitalmarktes in der Bundesrepublik - Ursachen, Auswirkungen, Lösungsmöglichkeiten. Tübingen.

Wetzel, W.E. (1983): Angels and Informal Risk Capital. In: Sloan Management Review (24), 23-34.

Wetzel, W.E. (1987): The Informal Venture Capital Market: Aspects of Scale and Market Efficiency. In: Vesper, K.H. et al. (Eds.): Frontiers of Entrepreneurship Research 1987. Wellesley, Mass., 412-428.

Wupperfeld U. (1994): Strategien und Management von Beteiligungsgesellschaften im deutschen Seed-Capital-Markt - Ergebnisse einer empirischen Untersuchung von 33 Beteiligungsgesellschaften und Banken. Working paper. ISI: Karlsruhe.

Wupperfeld U. (1996): Management und Rahmenbedingungen von Beteiligungsgesellschaften auf dem deutschen Seed-Capital-Markt. Frankfurt am Main.

III. Regional Networks for Technology-Based Firms

Innovative Regional Development Concepts and Technology-Based Firms[1]

Knut Koschatzky

1.   Technological Innovation and Regional Development

There is a close link between innovations and spatial factors. On the one hand, technological inputs are an important determinant for the development of regions and the spatial division of labour (Stöhr 1986; Keeble/Wever 1987; Porter 1990; Oakey 1994). On the other, "a specific regional economic structure, which can be described as the simultaneous existence of sectoral specialization and function differentiation ... (forms) the framework for innovative regional developments" (Kilper et al. 1994: 29).

There are empirical studies from various countries (e.g. USA, Great Britain, Germany, Netherlands) on differences in regional innovation. These differences are particularly marked in the case of large agglomerations (and the regions surrounding them) and peripheral, rurally structured regions. An innovation deficit can also often be observed in monostructural industrial regions. There are various **theoetical approaches** which attempt to explain these innovation differences:

- According to the product life cycle hypothesis (e.g. Vernon 1966; cf. also Schätzl 1996), innovation activities, as a central element in the early phase of the product life cycle, can be expected particularly in agglomerations. Here, the necessary features of localisation and urbanisation economies can be found, including a high density of businesses, a regional market and a qualified workforce.

- The theory of the spatially functional distribution of labour (e.g. Bade 1979) comes to the conclusion that agglomerations, and particularly their peripheries, represent suitable locations for innovation activities, since these business func-

---

[1] Parts of this contribution are based on research kindly supported by Deutsche Forschungsgemeinschaft within the framework programme "Technological change and regional development in Europe".

tions have a higher locational output and tend to displace business functions with lower locational output from central locations.

- The concept of the spatial division of labour within multiple firm companies (e.g. Tödtling 1990) emphasizes that depending on their location requirements entrepreneurial functions are distributed among firms in different regions. High-ranking entrepreneurial functions, which include research and development, tend to be concentrated at the central sites of enterprises, which in turn are located in or near agglomerations due to the necessary density of contacts.

Depending on location conditions relevant to innovation and diffusion, and the extent to which the region is equipped with technology and innovation services, regional innovation actors are either more or less successful in generating and marketing innovations, thus securing and enhancing income and employment. This chapter examines the interrelationship between innovation in firms and the firms' regional environment. On this basis, concepts and areas of activity are then discussed for promoting regional innovation potentials, particularly from the standpoint of technology-based firms.

2.   Interrelations between Innovation in Firms and the Regional Environment

The innovative behaviour of firms is basically influenced by factors present in the firm itself, in its environment, by location conditions and by policy framework conditions. Correspondingly, the literature identifies firm structure factors (see for instance Ewers/Fritsch 1989), business strategies and the direct environment of the firm (e.g. Tödtling 1990; Maas 1990), locational factors (cf. Köhler 1989; Gornig et al. 1992) and technology policy, regional and structural policy (Meyer-Krahmer 1990) as areas of influence. The various determinants for innovation in firms are summarized in Figure 17, following Meyer-Krahmer/Gundrum (1995).

Figure 17: Determinants of innovation in firms

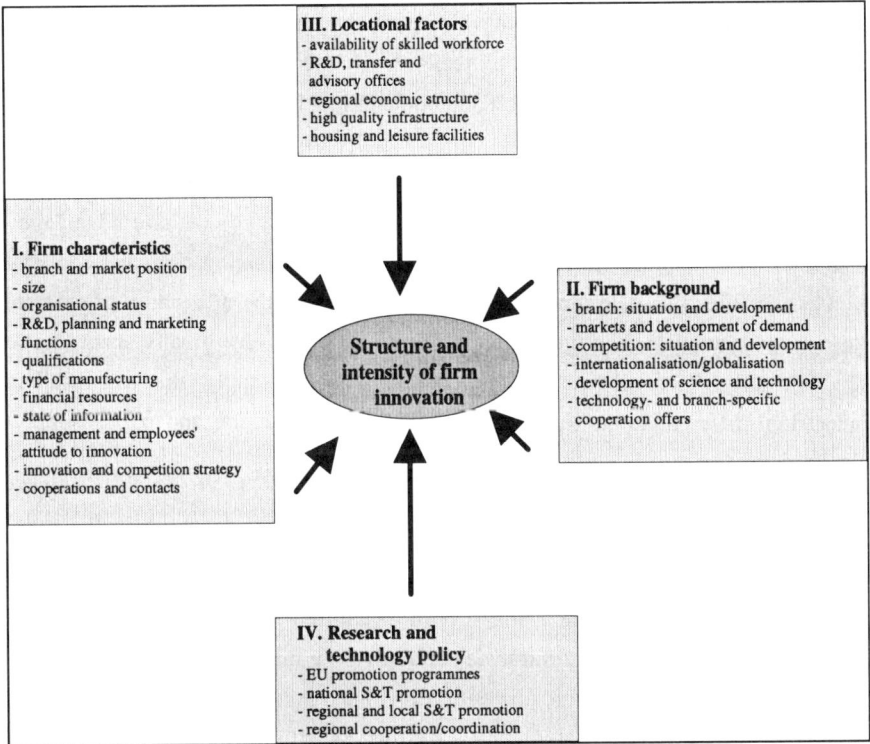

Source: Meyer-Krahmer/Gundrum 1995

Firm characteristics represent the primary area of influence on innovation in enterprises, with the firm size often regarded as a "sum variable"; this often also includes the strategic innovation resources (higher-value functions such as R&D and marketing, qualified personnel and information about new technologies and markets). The position of the firm is determined by its branch, by its organizational status and market position. These factors and the available resources define the possibilities for action which are expressed in specific forms of competitive, innovative and cooperative behaviour. An offensive strategy in these areas favours innovation within the firm. Financial restrictions, as well as personnel and information deficits, occur particularly in smaller firms. Organizational differentiation is ambivalent: it facilitates the systematic planning of R&D and information processing, but it may also hamper the flexibility of the enterprise; this flexibility is regarded as a particular advantage of smaller firms.

The **firm background** is closely related to spatial factors and the policy framework, which are treated separately for analytical reasons. The general background is determined by branch and market development, by the competitive situation, by techno-scientific development and by technology- and branch-specific cooperation. For the innovative enterprise, market and competition, branch-specific technological trends and cooperation opportunities are especially important. Product and process innovations aim at a corresponding market potential and are spurred by national and international competitive pressure (gaining or securing a competitive advantage). The relevant technology trends determine the direction of innovation, and cooperation possibilities facilitate innovations in firms (through the utlilisation of external knowledge potential). The opportunities for cooperation vary greatly according to the regionally-, technologically- and sectorally-specific supply; this can result in substantial differentiations (interrelation with spatial structure and policy support for cooperation).

Compared to the characteristics of the firm itself and its general background, the influence of **locational factors** is thought to be more limited. However, particular importance is attributed to the availability of a qualified workforce and the presence of technology-related research, transfer and advisory institutions as basic regional prerequisites for innovation. A diversified regional economic structure (broad sectoral spectrum), a central location and a well-developed infrastructure (fast interregional transport, telecommunication facilities) as well as high quality housing and leisure facilities are also considered as factors relevant for innovation. Whereas the significance of a favourable geographic situation as an influencing factor is decreasing as a result of continual improvements to the transport infrastructure, "soft" locational factors such as housing and leisure facilities and the quality of the natural environment are continually gaining influence due to their importance in attracting a skilled workforce.

The influence on innovation in enterprises of **research and technology policy** (R&T policy) at a European, national and regional level is regarded as secondary, compared to the characteristics of the firm itself. However, the relevance of R&T policy may be considered to have grown with the increasing differentiation and scope of promotion programmes (particularly at a European and regional level); the extent of its influence can be judged only with regard to specific branches, firm sizes or regions. Certain areas of technology, firm sizes and regions are definitely more strongly supported than others; this enhances the "policy awareness" of enterprises. The significance of policy at a regional level has increased; there are now numerous regional promotion programmes, which also include joint actions of the public sector with industry (e.g. technology and incubator centres). These cooperations can help to compensate regional supply deficits through synergy effects. In structurally weak regions, however, a technology-oriented regional policy requires complementary measures in other policy areas in order to create suitable locational conditions for innovation in firms.

Based on the innovation determinants described above, the following **hypotheses** can be formulated **on the interrelationship between innovation in firms and regional innovation**:

1. Innovation and technology adaptation in firms is favoured by high quality entrepreneurial functions, by appropriately qualified personnel (R&D, planning, marketing) and by organizational independence.

2. Innovation and technology adaptation in firms are fostered by a positive attitude within firms towards innovation, by an offensive innovation strategy and by close cooperation with external institutions (R&D, technology transfer and consulting).

3. Availability of qualified personnel on the labour market and the presence of R&D institutions, transfer and advisory units facilitate innovation and technology adaptation in firms.

4. R&T promotion specifically targeting certain regions, branches or firm sizes can support innovation and technology adaptation in enterprises, especially if it addresses the particular innovation barriers of firms (R&D personnel, innovation management, information on markets and technology).

5. Deficits in the facilities of structurally weak regions ("old" industrial regions or peripheral rural areas) can be combated by joint activities of the public sector and industry (exploiting synergy potentials in order to improve regional conditions for innovation).

6. Due to their lack of specialist capacities, small and medium-sized firms are less able to compensate regional deficits than are large enterprises (i.e., they are more strongly dependent on location factors); they often react by a particular form of

innovation behaviour: they tend to take over innovative production processes, rather than engage in innovation activities of their own.

3. Determinants of Regional Innovation Processes

The regional innovation potential is an important determinant for regional development (cf. Figure 18). It is understood as the characterization of all the factors that determine, or impede, the innovative performance of a region. It comprises the innovation actors of a region, i.e. primarily enterprises and research institutions, as well as the use made of the regional science and technology base and of the services supporting innovation and diffusion, such as offices for transfer and information. Thus when analysing regional innovation potentials, it is also necessary to examine the question of possibilities that have not been exploited so far. For this reason, the innovation potential includes not only technology and innovation-oriented firms, but also those with a low innovative performance, as well as the regional potential for firm foundations, since these represent starting-points for increasing the regional innovation potential. From the viewpoint of the regional economy, high innovation-related allocation gains can be expected if regional actors make particularly intensive use of the innovation potential.

Figure 18:   Regional innovation determinants

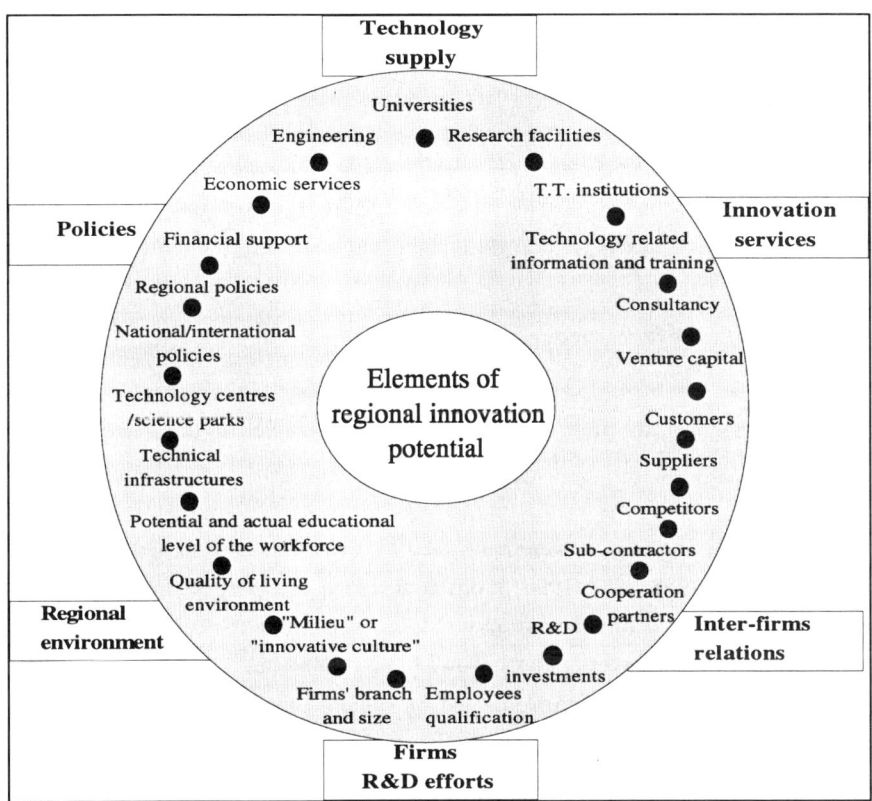

Source: Muller et al. 1994

Full exploitation of the regional innovation potential is realized by **activating re-**
gional innovation resources (e.g. entrepreneurial and private R&D, firm foundations, motivation of individual academic and industrial inventors) and by **interac-**
tions between innovation actors, i.e. research institutions, manufacturing and
service enterprises and transfer institutions. The interactions initiating a network
may occur as formal cooperations, but they may equally well take the form of informal contacts or the exchange of information. As well as technology and innovation-oriented measures, international and national policy frame conditions pertaining in the region also have an influence on the valorization of endogenous innovation resources. The regional environment also plays a part in the development of
regional innovation activities, for instance through the quality of the available infrastructure and the level of qualification of the workforce.

4. The Networking of Firms and the Development of Technology

As early as the beginning of the 1970s, relevant studies named the following prerequisites for successful technology development (cf. Rothwell et al. 1972):

- Accurate knowledge of user needs and conditions of use;

- Coordination of development, production and marketing activities within the firm;

- Making use of external information and consulting offers for science and technology;

- Basic research in the firm, linked with external research institutions (e.g. universities).

The prerequisites listed here were confirmed by numerous later studies (cf. among others Lundvall 1985, 1988, 1990; Wolff et al. 1994). All these studies emphasized the central importance of cooperation with users and with external sources of technological knowledge. Investigations showed that the internal R&D competences of innovative enterprises are complemented by regular or project-related links with universities, research laboratories, consultants and other firms. According to Freeman (1991), the particular problem in the innovation process lies in converting the various items of information into application-related knowledge for the development, production and marketing of new products and processes. Networks of firms with external partners in the manufacturing and service sectors can be useful in the gathering and processing of information (Tödtling 1994). The practically-oriented processing of specialist information can be supported by innovation transfer offices and advisory institutions. For the information to be used internally, however, there have to be appropriate interfaces within the firm.

"Classic" **networking relations** are between firms, their customers and their suppliers. Lundvall (1985) particularly emphasizes the importance of these links in complex technological innovations. Costly new developments can only be carried out in a demand related - and therefore marketable - form in close coordination with prospective users. Moreover, due to their technological complexity they require an increasing degree of specialization and differentiation on the part of enterprises, implying the necessity to outsource certain areas of production to qualified suppliers and to collaborate closely with them. The production process within the firm also changes as technological complexity increases, as does the division of labour vis-à-vis suppliers. Following the principle of flexible specialization, enterprises concentrate on the technological core areas associated with their particular know-how base ("lean production") and improve cooperation by adopting parallel work processes (e.g. development and production) and forming cross-departmental production groups with a broad spectrum of tasks. At the same time development, production and service tasks are outsourced to suppliers and service enterprises, a process which requires cooperation on a basis of equal partnership, with an open exchange of information (on innovation-oriented networking relations, cf. the next chapter in this reader).

Generally speaking, the vertical hierarchical production structure in the firm is replaced by horizontal cooperation involving external partners. Within these flexible networks consisting of the core enterprise and suppliers, technological developments are carried out in continuous interaction and communication. These networks are complemented by innovation-oriented services, e.g. in research and development, in information procurement and processing; as a consequence of the increasing specialization of manufacturing firms, the importance of these services is continuing to increase (Bade et al. 1989; Herden 1992).

5. Regional Technology and Innovation Concepts

5.1 Regional Technology Policy Framework

Regions are equipped with a specific permutation of production factors; these can only be considered to be optimally allocated if they are made the basis for regional technology and innovation promotion. **National technology policy** is not usually in a position to take regional problem situations adequately into account, since neither its goals nor its instruments are adapted to coping with regional peculiarities. A policy oriented towards the optimization of national innovation resources generally tends to increase regional disparities. Since important research institutions and industrial research laboratories are usually located in areas that are already economically favoured, public promotion of these locations tends to underline and reinforce existing regional disparities. Although financial transfers from a central government to regions in different stages of development may combat development and innovation barriers in individual instances, actions of this type will certainly cease to be suitable when measures are required which need to involve the education and training system, management competences or information and consulting aspects. Nevertheless, regional development can also be promoted by the central government. A **central regional policy** aims to reduce socio-economic disparities with reference to the national average. Infrastructural measures (e.g. involving transport, telecommunication and energy systems), regionally-differentiated investment grants and tax reductions can stimulate intraregional or external economic potentials and temporarily increase mobility of production factors oriented towards the region. In the long term, these measures can contribute to a stabilization in the range of interregional income disparities. Particularly in the less-developed regions of Europe, measures of this kind represent a (partial) compensation for the emigration of workers and the diminishing of economic activities. A regional corrective is supplied in larger federal countries by measures at state level (e.g. the German "Länder"). Due to the relatively small volume of the promotion funds, however, the extent of the impacts achieved by these Länder programmes is not generally comparable to the effects of national measures.

Through their enterprises and research institutions, regions are usually integrated into international and national technology diffusion and adaptation. There is little scope for regionally-specific technology development, although new technologies may definitely have a regional origin. However, because of the existence of regional focuses in technology, they can be made the starting-point for a regional development strategy. An **innovation and technology-oriented regional policy** is thus subject to restrictions which, although they may limit its scope for action, may also lead to promotion approaches that are sometimes very specific. When formulating measures, it is important to bear in mind that regional promotion activities are embedded in national and supranational science and technology policy, and cannot be considered in isolation from these levels. Enterprises, as well as research and development institutions, have the possibility of participating in national and supranational promotion programmes. Regional programmes must be oriented according to these higher-level promotion lines. Despite the existence of national and supranational (i.e. European) regional promotion, the extent of funding available for regionally-specific measures is limited. Promotion measures thus have to be planned and implemented, with the involvement of regional and local actors, in a form that will meet with a high degree of acceptance and will generate a correspondingly large regional impact.

5.2 Approaches for Regional Innovation Promotion

Basically speaking, the regional promotion of innovation and technology exploitation by firms can either use existing development potentials in the region as a starting-point (endogenous development), or can aim to attract new innovative enterprises from outside the region (mobility-oriented strategy). In view of the relatively small numbers of new enterprises coming in from outside, compared to the total population of firms in a region, the emphasis in innovation-oriented regional policy tends to be on **promoting the existing innovation potential** of the regional economy.

This type of promotion can adopt the following **approaches** (Meyer-Krahmer 1990):

- Increasing the region's innovation potential by motivating firms in the region to engage in innovation activities, and by stimulating technology-oriented firm foundations as "spin-offs" from existing enterprises and R&D institutions.

- Improving the regional conditions for innovation, especially innovation services and the relevant infrastructure (R&D institutions, information and advisory institutions, supply of capital and venture capital, facilities for training and further qualification).

- Stimulating existing firms (particularly small and medium-sized enterprises) to intensify their innovation activities.

These approaches are described in more detail in Figure 19.

Figure 19:   Approaches to the promotion of regional innovation potentials

Promotion of regional innovation and economic development

Attracting new technology-based firms (from outside the region)

Increasing the invention and adaption capability of firms already located in the region

Foundation of technology-based firms derived from existing potential

Increasing the attractiveness of the region by
- Formation of a technology region (image)
- Improvement of innovation-oriented infrastructure
- Improvement of "soft" locational factors (quality of living and leisure facilities)

Motivation and improvement of innovation conditions through
- Formulation of a demand oriented regional vision
- "Public-private partnership"
- Training and education facilities (professional qualification)
- R+D institutions and technology transfer
- Innovation services (information communication, consulting, cooperation)
- R+D cooperations
- Building up innovation, production and service networks
- selective support of industry and planning of commercial/industrial sites

Motivation and improvement of start-up conditions by
- Technology transfer
- Advice on firm foundation and management assistance
- Provision of (venture) capital
- Availability of suitable sites and infrastructure facilities (technology centres)

Mobility-oriented strategy   Endogenous development strategy

Source: Koschatzky et al. 1995

Measures should concentrate on building up the **innovation-relevant infrastructure** and **innovation services**; existing institutions in the region should be used as a basis for this and their strengths should be expanded specifically to meet regional demand. In particular, **technology and incubator centres** should be mentioned in this context (cf. also the following chapter on technology and incubator centres). In these are concentrated new, innovative enterprises which may be expected to generate positive effects in the region (cf. the bibliographical overview in Groß 1994). Technology and incubator centres should work in close cooperation with existing research institutions and technology-based firms in the region; structurally weak regions are disadvantaged here. Regional development strategy should aim at a balanced distribution of small and large enterprises in various different sectors and encourage networking among enterprises in R&D cooperations, supplier relations etc., because **production, service and information networks** favour the development of the regional economy and act as a catalyst in the exploitation of regional innovation potential. When doing this, regional contact networks should be used and particular emphasis should be given to regional identity. It is important to **formulate a regional model**, in which all the actors participate and bear joint responsibility (e.g. a "Technology Region"). This fixes the framework data for regional and technological development, assists actors in policy, research and industry, and the public as a whole, in their activities, and links all their activities to a common aim. This "public-private partnership" enables cooperative innovations to be realized which make consistent use of the region's resources (Fritsch 1990). As well as these measures with an economic orientation, it is necessary to aim to improve "soft" location factors (quality of housing and leisure facilities and the natural environment).

The main approaches to **promoting firm foundations** are: to ensure a supply of venture capital and business investment capital, not only for the start-up phase but also for the production and marketing phase of new technology-based firms; technology transfer; management consulting; supplying suitable industrial premises and infrastructure institutions. According to Keeble (1988), a particularly marked influence on regional development can be expected if new technology-based firms (NTBFs) enhance the competitiveness of established firms by developing new process technologies. Investigations on the influence of NTBFs on the regional employment market show that although these firms do create jobs, these do not keep pace with the jobs that are disappearing in traditional industry at the same time (Alaluf/Vanheerswynghels 1988). However, NTBFs can play a special integrative role, both at a regional and local level, between existing large enterprises, research and training institutions and public administration.

Research on firm founders' choice of locations has shown that the founding of a new enterprise usually takes place in the founder's regional environment (Fritsch 1990a: 243). The founder has most knowledge about this regional environment and here he also has his most intensive social contacts, which he can make use of when starting-up his firm. Analysis of the locational behaviour of founders supported by the "Eigenkapitalhilfeprogramm" (equity assistance programme, EKH) in Baden-Württemberg has shown that 71.9 percent of founders set up their own firm in the community where they themselves live (Schmude 1994: 79). This brings to light a **low willingness on the part of firm founders to be geographically mobile**, which underlines the important part played by firm foundations in mobilizing endogenous innovation potentials.

In Germany, most foundations of technology-based firms take place in central urban areas or the highly-populated areas of large agglomerations (Fritsch 1994: 15).

However, seen in relative terms - i.e. per 1,000 employees - there are more foundations in the outskirts of these regions ("suburbanization" effects), especially in regions which are quite close to cities and have a well-developed research infrastructure (Nerlinger/Berger 1995: 26). From this spatial pattern, it can be deduced that
regional environment plays an important role in the foundation activities of new
technology-based firms, for instance through the existence of incubator institutions
and the research infrastructure. On the other hand, analyses of the determinants of
innovation activities in small and medium-sized enterprises show that for successful
innovation internal factors, such as e.g. marketing organization and innovation
management, are more important for these firms than the regional environment
(Pfirrmann 1994: 52). Nevertheless, particularly for small and medium-sized firms
the importance of the part played in business success by the availability of information and advisory services in the region, and by integration into regional innovation
networks, should not be underestimated (Herden 1992).

As well as the tried and tested instruments that have proved their worth in numerous
regional development concepts, additional measures should be used to improve the
innovation capability of firms and their possibilities for dealing with technological
structural change. These include support for innovation management in firms, and
guidance on the organization of production and marketing (by consulting or training, for instance). The implication is not only that national technology policy should
reduce its overriding preoccupation with technology *per se*, but also that in regional
innovation policy more consideration needs to be given to factors such as information, management, the development of personnel and organization in an **integrated
innovation strategy**, in order to secure national and regional competitiveness in the
long term (Meyer-Krahmer 1993; Meyer-Krahmer/Gundrum 1995). The innovation
strategies that will be most successful in the future are those in which regional development is based on the conservation of resources, and those which aim to promote regionally-integrated production cycles with a high proportion of recycled
production inputs and longer-lasting products.

In the following section, the example of the Rhine-Main region is used to illustrate
individual economic and regional policy approaches which can contribute to improving regional innovation conditions.

5.3 Technology-Oriented Development Strategies for the Rhine-Main Region

The Rhine-Main region (cf. Figure 20), containing the cities of Frankfurt am Main, Darmstadt, Offenbach am Main, Wiesbaden, Mainz and Aschaffenburg, is Germany's largest finance and service centre. Leading German credit institutions and insurance companies have their headquarters in the region. Frankfurt, with its airport, is a nodal point for transport in Germany and Europe. The region is characterized by its numerous suppliers of business services. In 1987, 64.6 percent of the region's employees were working in the tertiary sector (compared with 47 % in Stuttgart and its surrounding area and 59.6 % in greater Munich). In 1990, the per capita income (gross national product per inhabitant) exceeded the national average by 43 percent (Koschatzky et al. 1993a). Despite these favourable economic conditions, the region has a structural deficit: its industrial base is small, and over the last two decades it has declined further. In 1987, only 13.4 percent of all jobs in the region were in the manufacturing sector. Since a large proportion of suppliers of regionally-oriented services are dependent on industrial demand in the region, any further decline in the importance of industry would be bound to have consequences for the service sector, too.

Against this background, the Fraunhofer Institute for Systems and Innovation Research (ISI), at the request of the Umlandverband Frankfurt and the Wirtschaftsförderung Frankfurt GmbH, elaborated a development concept for the region with the particular aim of strengthening enterprises and industrial sectors oriented towards high-tech. As a first step, a product list was drawn up (Koschatzky et al. 1992). This served as a basis for gathering empirical data on the high-tech enterprises and research institutions already located in the region. Backed up by interviews with industrial managers and policy-makers, a regional analysis of strengths and weaknesses was then performed. This formed the starting-point for the derivation of recommendations for regional action. In order to illustrate the possibilities for designing regional technology and innovation development concepts, the following section presents five **cross-cutting fields of action** which constitute the most important elements of the development plan elaborated for the Rhine-Main region with the aim of improving allocation conditions for technology-based firms (Koschatzky et al. 1993b).

Figure 20:   The Rhine-Main Region

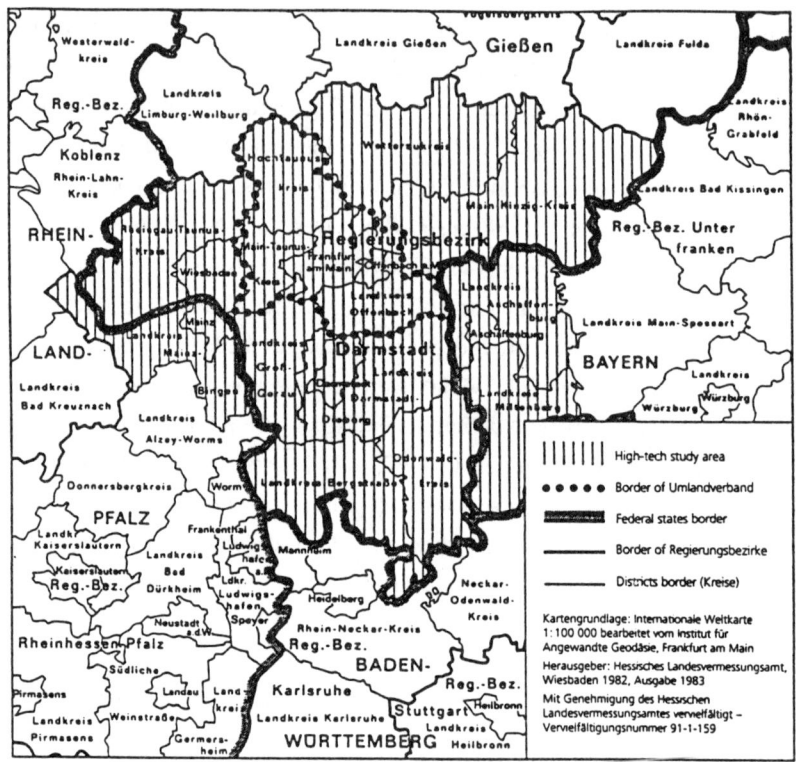

A technology-based model for the region

In the Rhine-Main region, 2,303 enterprises in the area of high technology (high-tech) were identified (Koschatzky/Kulicke 1994). Of these, 926 concentrated their activities mainly in manufacturing, 707 in wholesaling, service and training, and 670 in consulting and engineering services. From these figures, it is clear that the field of high-tech in this region is also predominantly a service field (i.e., tertiary). At the time of the survey, policy frame conditions in the part of the region belonging to Hesse were not oriented towards prioritising the promotion of the manufacturing sector. This, combined with the bottlenecks, particularly in the greater Frankfurt area, in availability of industrial sites, and the increasing allocation competition from Eastern European countries, created a danger that the industrial basis of the region would atrophy further. Due to the high degree of networking of many high-tech firms with other production and service enterprises, it is absolutely essential to keep the industrial element above a certain minimum in order to secure employment in the manufacturing and service sectors. The competitiveness of a region depends decisively on the development of technology and on its potential of innovating enterprises. However at the same time, technology development also determines industrial application and production. Therefore, concentrating on services does not suffice as a strategy to maintain and strengthen regional competitiveness. In order to safeguard and strengthen the technology-oriented industrial base in the Rhine-Main area, an economic policy model is needed for the region as a whole which emphasizes the importance of industrial firms for regional development, fixes promotion policy goals and areas of action, and in this way provides all decision-makers in industry, policy and administration with action guidelines that are unified for the whole Rhine-Main Region. This model should place emphasis on those functions and technologies in which the region is strong. These include research and development, innovation services (from pattern-making over information brokerage to financing), engineering, specialized (high-tech) manufacturing and the areas of marketing, selling, service, training. However, a sustained effect on economic development can only be achieved if the steering and decision centres necessary to these functions are located within the region.

Mixture of production and service activities on industrial sites

Through improved site management, the supply of commercial premises available in some areas of the region (the Frankfurt-Offenbach urban area) has improved. Nevertheless, there are still site bottlenecks, particularly in the urban centres. To counter this bottleneck situation, compact utilisation concepts are needed which can help to maximize the effective use of space. In addition, steps should be taken to ensure that sites previously in commercial/industrial use are not put to other uses, at least in situations where there is either an additional demand for commercial sites, or where the local business structure would make small-scale mixed-use sites combining production and service activities a possibility. The provision of sites of this kind would satisfy the specific requirements of technology-based firms for intense networking not only with other enterprises, but also with service suppliers and research institutions. It would not be a question of setting-up technology parks, but of creating acceptable location conditions in industrial/commercial sites that have already demonstrated their suitability for this group of enterprises (e.g. with regard to the variety of branches represented there, the availability of a qualified workforce, accessibility and use specifications. In mixed industrial/commercial estates of this kind, new forms of cooperation between firms could also be tried out, involving e.g. a minimization of transport processes between firms located there, or the creation of closed materials cycles, e.g. by inter-firm exploitation of (waste) heat or inter-firm waste recycling (e.g. eco-industrial parks).

Intensification of regional cooperation and technology marketing

The increasing internationalization of production can lead to a weakening of locational ties, and a consequent intensification of competition between industrial locations. This competition is particularly in evidence between technology regions. As a result of the increasing science base in technology development, i.e. the necessity for close cooperation with scientific institutions in the innovation process, location requirements arise which often can no longer be satisfied within one single city or community. In order to be able to continue to compete with other technology regions, the Rhine-Main Region has to present a unified picture of its technological competencies and market its allocation qualities accordingly. In order to achieve this, it is necessary to elaborate a joint technology concept for the region, since the economic strength of the cities and communities can only be secured by regional cooperation. The aim should be for cities and communities to cooperate in the declaration of industrial and commercial sites and to elaborate a key for the distribution of trade tax revenues. With the founding of the society "Wirtschaftsförderung Region Frankfurt/Rhein-Main", a first important step has now been taken towards coordinated promotion of the regional economy.

Authorisation procedures

One bottleneck repeatedly referred to in interviews with firms is the length of time required for the various authorization procedures in the part of the region belonging to Hesse (for instance with regard to building, immission and waste regulations). This leads to uncertainties for enterprises, with consequent disadvantages in competition (if competitors are able to start-up their new production units faster) and entails higher costs for firms compared with other locations. Since the main bottlenecks actually lie in the understaffing of the licensing authorities and in the procedures themselves, external institutions (for instance the TÜV) or engineering bureaux should be able to take on some of the tasks involved in licensing and inspection procedures. It should also be considered whether it might not be possible to replace the linear sequence of applications to the various authorising bodies by new organizational models, such as e.g. parallel procedures.

Research infrastructure

The region has 112 research establishments and institutes that carry out technology-based research. These are located in three universities and in seven "Fachhochschulen" (colleges of higher professional training) and non-university research institutions. Thus the Rhine-Main Region possesses a broad techno-scientific research base. Nevertheless, a repeated complaint was that Frankfurt University in particular was not sufficiently oriented towards technology and the natural sciences, in view of the technology profile of the region. As well as addressing the question whether certain specialist areas should be expanded or set up, or whether there is a need for more applied industrial research in the non-university sector, more intensive cooperation could also be envisaged between the universities. Due to the increasing networking of individual technologies (e.g. biotechnology/chemistry with microelectronics, metrology and materials sciences), and also to the necessity of including social policy considerations in technology development (humanization of technology, research on the impacts of technologies in the early stages of technogenesis) it would be necessary, as well as initiating cooperation in specialist areas between the three universities in the region (e.g. in chemistry, informatics) to bring in the social science competences of the J. W. Goethe-Universität, Frankfurt. In order to initiate cooperation over and above the present level, appropriate projects are required, but also a certain amount of start-up financing in order to enhance willingness to cooperate and reduce the costs of the ensuing process of institutional learning. Promising approaches can be observed at the TH Darmstadt (Darmstadt university of technology), where, in the Centre for Interdisciplinary Technology Research, technical aspects are considered in conjunction with societal policy and social science aspects of technology development.

These five fields of action were described in order to illustrate the possible content of regional concepts for technology development. In the detailed formulation of proposals, which should be carried out in close cooperation with responsible actors in the region, an aspect which should not be neglected is the way in which measures are to be implemented. This applies particularly to actions that require a regional and societal consensus. For instance, a regional initiative of the kind described for the Rhine-Main Region can only "come to life" if its aims are clear to all partici-pants, and if all those responsible for its realization are in agreement that, in a situa-tion of increasingly fierce technological and regional competition, the position of the region can only be maintained by ensuring a mixture of manufacturing and service enterprises.

6. Implications for Regional Technology and Innovation Concepts

Regions that are oriented towards technology and world markets are increasingly being drawn into an international location competition. New production and logistic concepts, as well as the use of information and communication technologies, are reducing the locational links of large enterprises in particular; the formation of geo-graphically dispersed, trans-sectoral networks means that the various business units of an enterprise may be sited at different locations, the specific territorial conditions required being dependent in each case on the business function involved. For indi-vidual regions, these developments imply that

- exogenous potentials for regional development, for instance by the allocation of new firms, become less significant, and that

- regions have to make more efficient use of their endogenous innovation poten-tials, exploiting them more fully, and have to adopt new, innovation-supportive approaches, in order to safeguard and enhance their employment and income levels.

As well the activation of endogenous innovation resources (e.g. by firm start-ups and the expansion of regional innovation services), support for an information, co-operation, production and service network is an important field of action for tech-nology- and innovation-oriented regional policy. Here, regions - or their policy agents - have decided advantages over national promotion administration. They are acquainted with the region and its actors, its cultural particularities and its special regional bottleneck factors. This knowledge does not, however, remove the need for a realistic, reliable analysis of regional strengths and weaknesses and, based on this, the formulation of recommendations for action. Before entering into a policy discussion on the future promotion of technology and innovation, regional actors must realize that the successful implementation of proposed measures is only possible if all those involved are prepared to cooperate. In order to arrive at this point, and to combat attitudes of self-interest, reluctance to make contact and rejection of technology, **new forms of cooperation** are essential. These include consensus-forming dialogues with all the participants in a regional development strategy (for instance in regional conferences, such as the ones introduced as a regional policy instrument in North Rhine-Westphalia; cf. Fürst 1994) and also the involvement of the general public in this dialogue.

Generally speaking, confidence-enhancing measures and new forms of cooperation should aim to initiate a regional discussion process between the general public and the responsible actors, in which various development possibilities for the region, their consequences for the population, and possible fields of action, can be discussed. Provided a common model has been evolved, and provided there is a high degree of acceptance for promotion measures, a contribution can be made to developing the region and strengthening its economic basis, even with limited funding. In parallel with these regional promotion measures, regional enterprises and research institutions should participate in national and supranational promotion programmes, in order to secure access to international technology and management knowledge. Thus a technology- and innovation-oriented regional policy is to be understood not only as an instrument for allocating regional resources, but also as a means of activating regional innovation potentials, while at the same time making judicious use of supranational, national and regional funding.

7. Bibliography

Alaluf, M./Vanheerswynghels, A. (1988): Local Employment, Training, Structures and New Technologies in Traditional Industrial Regions: European comparisons, in: Aydalot, Ph./Keeble, D. (Eds.): High-Technology Industry and Innovative Environments. London, New York.

Bade, F.-J. (1979): Funktionale Aspekte der regionalen Wirtschaftsstruktur. In: Raumforschung und Raumordnung (37), 253-268.

Bade, F.-J./Middelmann, U./Schüler, M. (1989): Expansion und regionale Ausbreitung der Dienstleistungen. ILS-Schriften 42. Dortmund.

Ewers, H.-J./Fritsch, M. (1989): Die räumliche Verbreitung von computergestützten Techniken in der Bundesrepublik Deutschland. In: Bröcker, J. et al.: Regionale Beschäftigung und Technologieentwicklung. Berlin.

Freeman, C. (1991): Networks of Innovators: A synthesis of research issues. In: Research Policy (5), 499-514.

Fritsch, M. (1990): Technologieförderung als regionalpolitische Strategie? In: Raumforschung und Raumordnung (48), 117-123.

Fritsch, M. (1990a): Zur Bedeutung des kleinbetrieblichen Sektors für die Regionalpolitik. In: Berger, J./Domeyer, V./Funder, M. (Eds.): Kleinbetriebe im wirtschaftlichen Wandel. Frankfurt, New York, 241-268.

Fritsch, M. (1994): New Firms and Regional Employment Change. Paper prepared for presentation at the 34th European Congress of the Regional Science Association, Groningen, 19-26 August 1994. Freiberg.

Fürst, D. (1994): Regionalkonferenzen zwischen offenen Netzwerken und fester Institutionalisierung. In: Raumforschung und Raumordnung (52), 184-192.

Gornig, M./Schultz, E./Einem, E. von et al. (1992): Mittel- und langfristige Entwicklungsperspektiven für Stadtregionen angesichts veränderter Rahmenbedingungen. Expertise for the Bundesministerium für Raumordnung, Bauwesen und Städtebau. Bremen, Karlsruhe.

Groß, B. (Hrsg.) (1994): Internationale Bibliographie zu Innovationszentren. ADT-Focus, Vol. 6. Berlin.

Herden, R. (1992): Technologieorientierte Außenbeziehungen im betrieblichen Innovationsmanagement. Heidelberg.

Keeble, D. (1988): High-Technology Industry and Local Environments in the United Kingdom. In: Aydalot, Ph./Keeble, D. (Eds.): High-Technology Industry and Innovative Environments. London, New York.

Keeble, D./Wever, K. (Eds.) (1987): Firm and Regional Development in Europe. London.

Kilper, H./Latniak, E./Rehfeld, D. et al. (1994): Das Ruhrgebiet im Umbruch. Strategien regionaler Verflechtung. Schriften des Instituts Arbeit und Technik (8). Opladen.

Köhler, S. (1989): Diffusion und räumliche Wirkungen neuer Informations- und Kommunikationstechniken in der Bundesrepublik Deutschland. Working Paper of Institut für Städtebau und Landesplanung der Universität Karlsruhe. Karlsruhe.

Koschatzky, K./Grupp, H./Gundrum, U./Hinze, S./Kuntze, U. (1992): High-Tech-Unternehmen in der Region Rhein Main. Grundlagenstudie. Frankfurt, Karlsruhe.

Koschatzky, K./Breiner, S./Bördlein, R. et al. (1993a): Technologieprofil der Region Rhein Main. Hauptstudie, Teil I zum Projekt High-Tech-Unternehmen in der Region Rhein Main im Auftrag des Umlandverbandes Frankfurt und der Wirtschaftsförderung Frankfurt GmbH. Frankfurt, Karlsruhe.

Koschatzky, K./Breiner, S./Gundrum, U. et al. (1993b): Standortvoraussetzungen und Fördermaßnahmen für High-Tech-Unternehmen in der Region Rhein Main. Hauptstudie, Teil II zum Projekt High-Tech-Unternehmen in der Region Rhein Main im Auftrag des Umlandverbandes Frankfurt und der Wirtschaftsförderung Frankfurt GmbH. Frankfurt, Karlsruhe.

Koschatzky, K./Kulicke, M. (1994): Policies towards Technology-Based Companies in a Regional Context. In: Gonda, K./Sakauchi, F./Higgins, T. (Eds.): Regionalization of Science and Technology Resources in the Context of Globalisation. Industrial Research Center of Japan: Tokyo, 143-164.

Koschatzky, K./Gundrum, U./Muller, E. (1995): Regionale Innovations- und Technologieförderung. Ansatzpunkte für die Nutzung regionaler Innovationspotentiale. Working paper. ISI: Karlsruhe.

Lundvall, B. (1985): Product Innovation and User-Producer Interaction. Aalborg.

Lundvall, B. (1988): Innovation as an interactive process: from user-producer interaction to the national system of innovation. In: Dosi, G./Freeman, C./Nelson, R./Silverberg, G. (Eds.): Technical Change and Economic Theory. London.

Lundvall, B. (1990): From Technology as a Productive Factor and an Interactive Process. Paper at the Montreal Conference on Network Innovators. Montreal.

Maas, C. (1990): Determinanten betrieblichen Innovationsverhaltens - Theorie und Empirie. Berlin.

Meyer-Krahmer, F. (1990): Innovationsorientierte Regionalpolitik: Ansatz, Instrumente, Grenzen. In: Gramatzki, H.-E. et al. (Eds.): Wissenschaft, Technik und Arbeit: Innovationen in Ost und West. Kassel.

Meyer-Krahmer, F. (1993): Welche Technologiepolitik braucht der Standort Deutschland? Wirtschaftsdienst XI, 559-563.

Meyer-Krahmer, F./Gundrum, U. (1995): Innovationsförderung im ländlichen Raum: Raumforschung und Raumordnung (53), 177-185.

Muller, E./Gundrum, U./Koschatzky, K. (1994): Horizontal Review of Regional Innovation Capabilities. Final Report. ISI: Karlsruhe.

Nerlinger, E./Berger, G. (1995): Regionale Verteilung technologieorientierter Unternehmensgründungen. Discussion Paper Series, ZEW Mannheim, Nr. 95-23. Mannheim.

Oakey, R.P. (1994): High Technology Small Firms and Regional Development in the United Kingdom: Some Conceptual Observations. Paper presented at the Berlin Symposium, Oct. 17 and 18. Manchester.

Pfirrmann, O. (1994): The Geography of Innovation in Small and Medium-Sized Firms in West Germany. In: Small Business Economies (6), 41-54.

Porter, M.E. (1990): The Competitive Advantage of Nations. London.

Rothwell, R. et al. (1972): SAPPHO Updated. In: Research Policy (3), 259-291

Schätzl, L. (1996): Wirtschaftsgeographie 1: Theorie. 6th Edition. Paderborn.

Schmude, J. (1994): Geförderte Unternehmensgründungen in Baden-Württemberg. Erdkundliches Wissen, Heft 114. Stuttgart.

Stöhr, W. (1986): Territorial Innovation Complexes. Interdisziplinäres Institut für Raumordnung, Stadt- und Regionalentwicklung, IIR-Discussion 28. Vienna.

Tödtling, F. (1990): Räumliche Differenzierung betrieblicher Innovation - Erklärungsansätze und empirische Befunde für österreichische Regionen. Berlin.

Tödtling, F. (1994): Regional networks of high-technology firms - the case of the Greater Boston region. In: Technovation (14), 323-343.

Vernon, R. (1966): International Investment and International Trade in the Product Cycle. In: Quarterly Journal of Economics (80), 190-207.

Wolff, H./Becher, G./Delpho, H. et al. (1994): FuE-Kooperation von kleinen und mittleren Unternehmen. Heidelberg.

Innovation Networks for Small Enterprises[1]

Knut Koschatzky, Uwe Gundrum

1. Innovation Networks: Why They are Formed, How They are Structured

Enterprises generally network closely with their environment. The economic importance of these linkages, which may take the form of production and service alliances, research cooperations or innovation networks[2], has increased greatly over the past decades. This development has been triggered by two main factors: the global cost and price competition between firms and the rapidity of innovations in information and communication technology, enabling internal and external cooperations to take place which could not have been envisaged a few years ago. These include, for instance, electronic publishing and the worldwide exchange of production and management data.

"Classical" **networking relationships** have always existed between firms and their suppliers and customers. However, these have always been vertical and hierarchical in character, with the companies retaining great scope of production. Even the increasing fragmentation of production tasks into individual steps, that commenced at the end of the nineteenth century, rendered possible by the mechanization of pro-

---

1 Parts of this paper have been presented at the RESTPOR '96 conference on "Global Comparison of Regional RTD & Innovation Strategies for Development and Cohesion", Brussels, 19-21 September 1996.

2 The term "network" is understood to cover a wide variety of formal and informal contacts and contractual relationships, which may extend from information procurement, through financial aspects, to supply and sales relations and strategic alliances. Industrial networks are based on mutual exchange within a system of interdependent, dynamic relationships (Gemünden 1990; Powell 1990). Each participating partner is mutually dependent upon resources controlled by the other, so that certain goals only become attainable when their divided resources are combined (Willms et al. 1994). Through networks with a judiciously planned division of tasks between partners in the production and service sectors, optimal use can be made of the development potentials present in a region, and synergy effects can be exploited, although an intensification of these networking relationships is usually accompanied by an increased intensification of communication and transport.

duction, (tayloristic manufacturing concept), took place within the firm and only very gradually led to the outsourcing of manufacturing processes to other firms.

New production concepts only began to appear in the nineteen-seventies, following Japan's entry into the world market and the ensuing shift in global economic equilibrium, which brought with it structural changes in competition regarding cost, quality, price and innovation. On the one hand, the new production concepts targeted the flexibilization of management and production processes and the integration of all company activities through new information and communication technologies (CIM-technologies). On the other hand, these concepts aimed to cut down the numbers of employees and enhance the skills and qualifications of the remaining workforce. Following the concept of flexible specialization, firms are now concentrating on their technological core areas based on special know-how ("lean-production") and on improving internal cooperation through re-engineering by parallel work processes (for instance, close interaction between development and production) and by cross-departmental production groups. At the same time, development, production and service tasks are being outsourced to suppliers and service companies, a prerequisite for this being a relationship of cooperative partnership with an open exchange of information. Suppliers are closely linked with this process through their direct involvement in the production processes of the recipient firm. For the core enterprise, an important aim is to reduce their number of suppliers by the creation of "system suppliers" who deliver complete systems or modules. One consequence of lean production has been the increase in "just-in-time" supply relations, leading to a redistribution of costs between firms and the rest of society (reduced inventory carrying costs, accompanied by an increased frequency of transport and its impacts on the transport infrastructure and the environment).

Another development is that the importance of **services** in the context of industrial production has markedly increased. This is reflected not only in the ongoing tertiarization of the economy, but also in the share of service functions within the manufacturing sector itself. Thus, approximately 40 percent of employees in the goods producing industries are concerned with services (Reichwald/Möslein 1995). The servicing of machines and equipment is just one instance of this service aspect of production; the construction of complete production plants (engineering) and the training of personnel by equipment manufacturers also belong to this area.

Figure 21:   Regional innovation networks

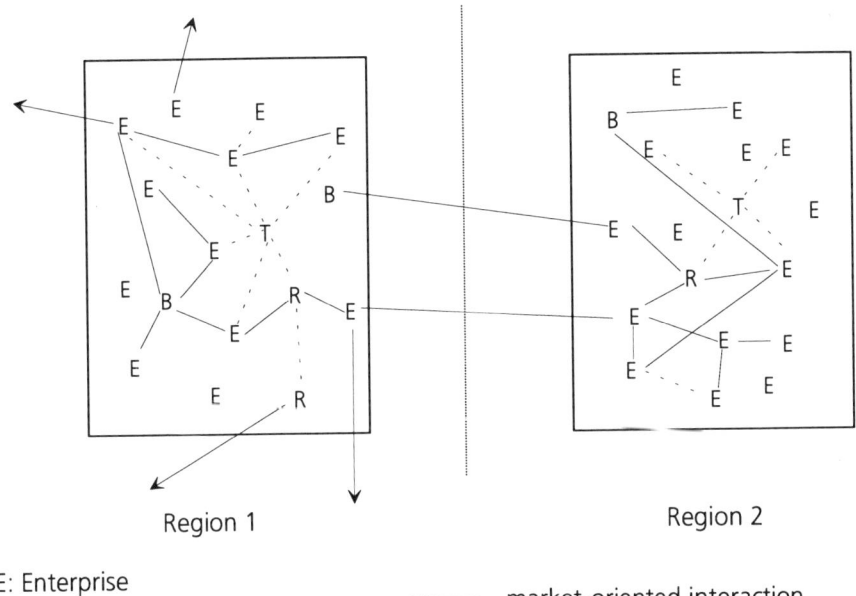

Region 1   Region 2

E: Enterprise
R: Research Institution   ——— market-oriented interaction
T: Transfer Office   - - - - non-market-oriented interaction
B: Bank

Regional cooperation networks act as catalysts in the exploitation of regional innovation potentials (Tödtling 1994). Innovation barriers within firms can be overcome through cooperation with other partners; information deficits can be removed
by the exchange of information within the network. In the ideal case, a network of
local, regional and supraregional relationships arises at a regional level in the form
of market-based and non-market-related interactions[3], characterized by contacts, the
exchange of information and formal and informal cooperations (cf. Figure 21). The
purpose of these relationships for the enterprises is reduction of their transaction
costs and the efficient buying-in of complementary research capabilities. This applies particularly in cases where the spectrum of potential partners' offers is known
and corresponds to the enterprise's own requirements, where the number of actors
exceeds a minimum threshold necessary for mutual cooperation, and where the

---

3   Market-oriented interactions are expressed as contractually agreed, financially remunerated R&D
cooperations between firms, or between firms and research institutions; non-market-oriented interactions may arise through the exchange of ideas, the passing-on of information and the transfer of (scientific) knowledge.

linking of competences promises a high degree of synergy on both sides. Cooperations may develop vertically (i.e. between partners at different levels relative to the market), horizontally (i.e. between partners at the same market level, e.g. competitors, enterprises in complementary areas of technology) and diagonally (i.e. between partners both differently and similarly placed relative to the market; these may be formed, for instance in order to jointly open up new sales potentials).

2. Basic Elements of Innovation Networks

The **network approach** was developed primarily in Scandinavia (cf. particularly Håkansson 1982, 1987 and 1989). It was initially applied to cooperations between firms formed with the specific aims of joint product development, production and distribution; it was later extended to include public organizations such as education and research institutions, industrial associations and regional bodies. The **basic elements** of networks are the players and their activities (development, combination, exchange and transformation of resources; cf. Håkansson 1987, 1989 and summary presentations in Herden 1992; Fritsch 1992). These activities in turn require resources for action, which may consist of the production factors labour, capital and raw materials/plant, as well as political-legislative and administrative basic potentials (e.g. regulations legitimized in policy, organizational steering capacity).

According to Håkansson/Johanson (1984), a network is characterized by the specific formation of **four linking characteristics**:

- Functional interdependencies: a concrete relationship system of actors, activities and resources;

- Power structures: influential potentials of actors who control activities and resources;

- Knowledge structures: activities and resources based on the experience and knowledge of former players;

- Time structures: the network is the product of experiences and investments in knowledge, relationships, behavioural patterns, etc.

Networks are also described as "resource coordination systems between market and hierarchy", i.e. the activities of the players are not coordinated by the price mechanism or by hierarchical dependence, but by concrete relationships of exchange within a specific network (Herden 1992). The absence of a formal hierarchy does not imply that these relationships of exchange are fundamentally symmetrical or in equilibrium; rather, in the field of industry in particular there are **dominant actors** with superior resources for action who are able to influence the formation of network relationships extensively in their own interests (e.g. the dependency relationship of suppliers and manufacturers); the decisive factor, however, is that the network as a whole is of use to all participants (on this aspect cf. Håkansson/Johanson 1993). The exchange relationships between the partners may relate to various **levels of action:** technology, planning, knowledge, socio-economic and legal arrangements. **Exchange relationships** with the transaction partners may be **direct and indirect** (e.g. customers, suppliers, complementary enterprises); indirect relationships also serve to access "external" resources beyond the actual partners and outside the network itself. Further distinctions are made regarding the **intensity and orientation of the exchange relationships**: There are very close relationships, for instance cooperation relationships between enterprises and their suppliers, and looser relationships to a wider circle of players with whom firms only "associate" from time to time. The transaction costs are higher in the close exchange relationships than in the looser ones; in the latter, exchange is possible with a broader environment. Exchange relationships naturally tend to be closer in the core area of networks than at its periphery; the most important transactions between partners take place at the core. However, the relatively "loose" relationships with the broader environment are also important. They increase the networks' flexibility and their openness to new frame conditions and requirements; they open up additional potentials for action.

This **ability to react and learn** through loose but widespread links to other players is particularly significant for innovative networks (on this, cf. Grabher 1992; 1993). This broad, extensive network structure comprising contacts of varying intensities brings with it a redundancy within the network, which reduces the degree of dependence of the individual partners (creation of buffer capacities). Successful networks are characterized by a high proportion of non-hierarchical cooperation relationships (Fritsch 1992): the absence of hierarchical structures is, in fact, a prerequisite for cooperation and "soft" interaction. Nevertheless, there are often players who function as **coordinators** within the networks, in the sense that they assemble information and undertake infrastructural tasks for the network (such as training or financing). These intermediaries may be private or public organizations.

Decisive for the stability and effectiveness of the network is the building-up of a mutual relationship of trust between the partners based on successful cooperations, which in turn are based on complementary resources, common aims and shared patterns or rules of behaviour (cf. Fritsch 1992). This **joint basis for action** facilitates close cooperation between the partners, dismantling individual insecurities and deficits in resources and opening up additional synergy potentials. The decreasing transaction costs and the generally improved results bring **cooperation advantages** for all the participants, thus enhancing the stability of the network. Moreover, it is often only by the judicious pooling of individual resources that complex problem constellations can be effectively faced and dealt with.

3.   Regional Innovation Networks of Small Enterprises

Various empirical studies (e.g. Perrin 1990; Wolff et al. 1994), have shown that locational proximity of the actors favours R&D cooperations and that the regional environment exerts an important influence on the formation of innovative networks. The concept of the "innovative milieu" developed by GREMI (e.g. Aydalot 1986; Camagni 1991; Maillat 1992; Quevit 1991), postulates that territorial dynamics depend upon the behaviour of all actors in the milieu. The **region** provides enterprises with a structural basis for their development (outsourcing of functions, local relations). The firms have an interest in their own integration into the milieu, and themselves enrich the milieu by their formation of territorial networks. These networks consist in market relations between firms, customers and suppliers and cooperations with research centres and universities, as well as cooperation with the training and education system and public administration. They are supported by regional cultures and identities, within which the individual actors are activated by a common (regional) goal. According to the GREMI argumentation, the **innovative milieu** can be described by **three** distinguishing **characteristics** (Koschatzky et al. 1995):

- a local framework, i.e. a geographical territory associated with homogeneous behaviour of the players;

- a functional logic, i.e. the actors of the milieu cooperate with the aim of developing innovations;

- "knowledge dynamics" which extend the actors' capabilities.

Although the regional socio-economic concept of the innovative milieu is frequently addressed in the literature, no quantitative foundation for this approach has yet been advanced, and the question of its transferability to regions with differing socio-cultural backgrounds has yet to be clarified. However, this concept does describe the framework which can favour the formation of regional innovation networks and it highlights the advantages of regional cooperation.

Whereas proponents of the innovative milieu emphasize the importance of the region for the formation of networks and as a structural basis for enterprises, the **theory of flexible specialization** places new industries and enterprises at the centre-point (Storper/Walker 1989; Storper 1992). According to this latter concept, the development of regions is dependent on the changing forms of organization of new industries. Growth industries are characterized by flexible production concepts, such as vertical disintegration, i.e. the outsourcing of production processing sequences to external partners. Around the firms there arises a network of new supply and sales structures which influence the image of the region. This theory does not regard the joint interests of firms as lying in a uniform regional identity but in competitive production based on the consistent application of new production, management and logistics concepts. The more intensively enterprises in the "new industrial districts" cultivate intraregional and interregional exchange relationships, i.e. the more they externalize their transactions, the stronger will be the exploitation of existing growth potentials. Regions with a high density of flexibly structured companies profit from intensive regional and supraregional production networks, which are often hierarchically organized, and through them they gain growth advantages over regions characterized by enterprises with "fordistic" production concepts.

Regarding **enterprises** in national and regional innovation systems, it emerges clearly from both these concepts that the requirements of research and development (R&D) in terms of performance and reactive capability are increasing all the time. This is attributable to the ever-shorter product life cycles, the globalization of production and procurement, the growing technology content of products and production and the resultant stepping-up of competition between firms. Whereas large enterprises can meet these challenges in their own R&D departments, most small and medium-sized firms (SMEs) are not in a position to fulfil the necessary conditions using their own resources. This is due to **several factors:**

- SMEs do not have a fully-developed internal R&D infrastructure; thus they have to rely on cooperations.

- SMEs generally have a low degree of organization and a decision structure involving only a few people. The decision-makers are tied up with day-to-day business. Information is only processed if it is relevant for matters requiring immediate decisions.

- Information procurement via the usual channels is not sufficient for SMEs. There is a need for "mediators" to lead the enterprises step by step, by personal contacts, towards the external R&D infrastructure and the exploitation potentials it offers.

According to the transaction costs approach (cf. Coase 1937; Picot 1981), networks can reduce costs within firms; thus under free market conditions, numerous networks of this kind would be expected to develop. Since, however, with the exception of supply and sales networks this is not the case in many regions, the non-functioning market mechanism (which may fail due to insufficient information, for instance) has to be substituted by state **intervention**. At a regional level, innovation and technology policy measures have to create frame conditions that enable SMEs to become embedded in innovation-oriented regional networks.

4. Examples of Regional Innovation Networks of Small Enterprises

The distinguishing characteristics of regional firm cooperations today lie in the **specifically-targeted division of labour** and in the **mutual interdependency of large, medium and small firms in the production and service sectors**. The networks of enterprises are no longer one-sided, in the sense of supplying the core en-

terprise, but consist of numerous cooperations between specialized partners in production and services. These modern networks of enterprises thus involve complex
exchange relationships between diverse partners on a horizontal and a diagonal
level. The development of the regional economy is stabilized by the variety of
branches and firm sizes, with new technologies and innovation services playing a
special role.

Baden-Württemberg is often named as an exemplary model of an heterogeneous
economic structure at a high technological level. Due to its lack of raw materials
and its unfavourable transport situation, this region was forced early on to concentrate on special manufacturing processes with a high value added (cf. Sabel et al.
1989; Scott/Storper 1990; Herrigel 1993). Out of this situation, a close-knit cooperative network has developed consisting of "Mittelstand" enterprises in the areas of
mechanical engineering, metrology, optics and electronics/electrotechnics on the
one hand, and on the other of large enterprises operating worldwide in mechanical
engineering and vehicle construction, electrotechnics and electronics. Due to this
combination of close cooperation with a high degree of specialization and the early
use of new manufacturing processes, the product programme was continuously developed and adapted to changing customer needs. A network of small and medium-
sized firms with flexible division of labour and a differentiated organizational
structure has arisen in **Northern Italy** (cf. Sforzi 1989; Lazerson 1993). This network tends to be horizontally structured as a whole, and is based upon long-term
cooperative relationships; it is supported by public promotion measures (e.g. industry and craft parks, regional marketing associations) and is, like Baden-Württemberg, a model example of an "Industrial District".

As well as these production-alliances, there are regional **service networks** in Germany such as the "Forschungs- und Transferverbund" (research and transfer association) in the Hagen region and **mixed cooperation networks** between communities, research institutions, chambers of commerce, banks and enterprises for the
promotion of regional technology and innovation, such as are found in the technology regions of Aachen and Karlsruhe. There are "mixed" industrial parks housing
enterprises, research and transfer institutions; there are local and regional development projects with a cross-sectoral orientation (e.g. the Emscher Park Internationale
Bauausstellung, IBA). Here, commercial premises, services for enterprises, housing
and leisure facilities are carefully combined, in order to initiate regional develop-

ment impulses on a broad front, while at the same time exploiting cooperation advantages and synergy effects. When planning industrial estates and expanding new information and communication technologies, firms in the production and services sectors can be specifically combined and networked with one another.

5. Measures for the Promotion of Innovation-Oriented Cooperations by Enterprises

Against the background of the increasing globalization of technology development and production, and in view of the necessity for making judicious use of regional development potentials for international competition, **regional innovation promotion has a triple task:**

- The activation and careful complementation of regional resources for the development and application of new technologies (regional innovation conditions);

- The coordination and interlinking of these resources in regional innovation networks, bringing in all the relevant actors in industry, science and policy;

- The integration of these regional networks into national and international clusters of technology development and production, by the creation of active interfaces and the promotion of supraregional cooperation.

Thus there is a need for **targeted initiatives at a regional level,** directed towards priority problem areas and bringing together regional players by actions that are project-based or measure-based. These initiatives have to continually secure the cooperation process by the provision of resources (personnel, finance, equipment) and by policy legitimation. They are mainly initiated and carried out by regional key figures (promoters) in politics, industry and science, but they may be supported by external specialist advice, regional coordination and decision-making committees and by the promotion measures of policy at higher levels.

From the viewpoint of small enterprises, information on the following aspects may have the effect of stimulating cooperations (Willms et al. 1994: 208ff.):

- Information on the spectrum covered by regional R&D institutions,

- Information about promotion possibilities,

- Information about possible cooperation partners,

- Information on technological developments and ways of solving problems in the enterprise itself,

- Consulting enterprises before engaging in projects,

- Initiation and mediation of topic-based and firm-centred cooperations.

The demand-oriented creation and extension of networks should be initiated by **industrially-oriented moderators** who have specialized technical qualifications, communicative competence, authoritative and well-accepted personalities and good relations with regional administration. These may be:

- Representatives of research institutes,

- Entrepreneurs,

- Public administrators.

If no networks have yet developed in the region, "public moderators" could be appointed, with the following tasks:

- Performing objective diagnoses for enterprises,

- Assisting in removing barriers,

- Convincing firms of the need for innovation planning in the enterprise,

- Overcoming the doubts of entrepreneurs.

These tasks need personal contact; contact should be given priority over technical information.

Moderators should initiate regional cooperation and spur the further development of the network. It is helpful to formulate a common objective, which might take the form of a joint project, for instance. As well as specifically building up innovation networks, the uncontrolled proliferation in the transfer landscape should be reduced to order (creation of a system with clearly defined competences and responsibilities). It would also be possible for heads of research institutes to function as specialized promoters with regard to networks. Their tasks could be:

- Active external representation of the specialist field,

- Bringing together potential addressees,

- Initiating networks for the development of joint research projects, e.g. in transdisciplinary technologies (communication technologies, new materials, environmental technology), in which large enterprises and SMEs (as potential suppliers) are brought together.

As well as these regional development measures, there are possibilities for **specific cooperation promotion** between firms and institutions in the innovation infrastructure (R&D, technology transfer, information and consulting, innovation financing and management, promotion of firm foundations). The following measures serve as an example for the promotion of the cooperative capability of regional enterprises and research institutes (cf. Gundrum/Walter 1995; Reiß/Koschatzky 1997):

- The promotion of joint innovation projects between firms, research institutions and other innovation actors in selected areas of technology ("joint innovation"),

- Assistance for small and medium-sized enterprises in participating in these projects,

- Expansion of contract research institutions at universities.

The selection of suitable areas of technology had to take account of the existing strengths of the region and a sufficient availability of transdisciplinary technologies (e.g. microelectronics, new materials, biotechnology); these constitute the basis for innovative problem-solving. The basic idea behind the **promotion of "joint innovation projects"**, a concept similar to the Small Business Innovation Research (SBIR) programme in the United States, is that the sequential and feedback character of innovations should be transferred to innovation promotion. Although only two phases are funded by regional governments according to this model (cf. Figure 22), the decision to promote an innovation project depends upon the realization chances of the whole project. Through this mechanism firms are forced to care about financing and marketing already at an early stage in the innovation process and to cooperate with different partners in order to bring the project to a successful end.

Figure 22: Promotion of joint innovation projects

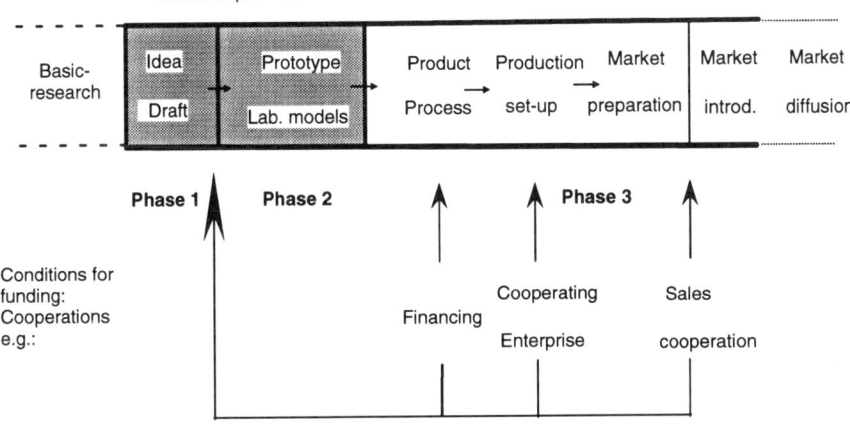

The following measures are designed to promote the **participation of small and medium-sized firms in these projects:**

- Information services, providing detailed information on the promoted projects and possible cooperation partners with respect to the different "phases" of innovation processes,

- Support of consulting on organizational, legal and technical questions relating to cooperation (free initial consultation, transparent offer),

- accompanying measures involving personnel and training, in order to improve cooperative capabilities (e.g. university graduates as "innovation assistants").

Contract research institutions at universities improve the transfer of research results to industrial application; their purpose should be to carry out practically-oriented, continuous research which the universities themselves are unable to perform in the course of their teaching and research commitments. The basic financing of the institutes is provided by public bodies (the regional government, communities and industrial enterprises); promotion associations or foundations can be set up for this purpose. However, the institutes have to finance themselves mainly by research and development for industry (similarly to the model of the Fraunhofer Society in Germany).

These proposals are complemented by measures for the support of cooperation involving personnel and training: university graduates as "innovation assistants" and the formation of "mixed teams" combining the firm's own specialists with external specialists improve conditions for the cooperative development of new products and processes in small and medium-sized firms (which do not usually have the necessary specialist personnel themselves). Both these measures promote cooperation between regional R&D institutions and small and medium-sized firms, and are jointly organized by state and industry. In parallel, training measures are recommended in communicative and social skills and transdisciplinary, non-compartmentalized modes of thought (integrated approach to technology and organization). Here, the training and education institutions would have to cooperate with the enterprises.

Another approach to the support of cooperation is through **regional transfer centres** such as those of the Steinbeis Foundation, opening up regional R&D potentials for small and medium-sized enterprises, and sectorally-oriented "cooperation exchanges" and cooperation catalogues. Another possibility is technology-oriented demonstration and training institutions such as CAD/CAM laboratories or CIM centres, which would support particularly small and medium-sized enterprises (including handicraft firms) in the introduction of new technologies; thus these also constitute examples of cooperative innovation support.

Another aspect of the promotion of innovation networks is the improvement of communication and the exchange of information among small and medium-sized firms. **Information and communication** between the various players in a (regional) innovation system are decisive prerequisites for the optimal exploitation of regional innovation resources and thus constitute important elements in regional concepts for the support of innovation networks. In the early 1970s, the importance of information and consulting for successful technology development was already being emphasized in empirical studies (e.g. Rothwell 1972). According to the transaction cost theory, the exchange of information between firms and their innovationally relevant environment reduces the development and transaction costs of the individual enterprises. Imai/Baba (1989) highlight the relevance of new information technologies for entrepreneurial networks. These networks have been supported and facilitated by the judicious use of new information technologies. The interrelationship between continually more complex software products, new technology applications by users and the increasing demand for services (information, consulting and technological servicing) is considered to favour the formation of new entrepreneurial networks (Koschatzky et al. 1993: 62).

In the advisory project "Regional Innovation and Technology Transfer Strategies and Infrastructures" (RITTS) Bremen, which was partially financed by the EU-Commission, ISI suggested together with other European experts to improve the networking and communication between firms and research, transfer and advisory institutions by means of an **online supported interactive communication forum** (cf. Schmidt 1995: 86). A survey, carried out during this project, of cooperation between enterprises and the involvement of regional research and transfer institutions in firms' innovation projects, indicated that not all the 60 enterprises interviewed were conducting an intensive search for partners and information. Smaller enterprises repeatedly remarked that the offer of information relevant to innovation and technology was too complex for them to be able to make a systematic search and that, due to their lack of staff capacities, it was not possible for them to gain a comprehensive overview of the services offered by the R&D infrastructure in Bremen. On the other hand, a survey of research institutions in Bremen revealed that these only very occasionally addressed SMEs in their regional environment or informed them of their research and consulting offer. In the RITTS project, it was concluded from this survey that a clear entry-point should be designated for the research and transfer offer in Bremen (in the sense of "One-Stop Shopping") and that opportunities should be provided for simple communication between the individual innovation players.

The proposed **electronic network** was intended to fulfil two main tasks:

1. To provide a pool of technical and policy information, in which the user could locate information about research programmes, promotion programmes, current research topics of institutes in Bremen, the activities and emphasis of local research institutions, conferences, seminars, qualification offers, etc.

2. To act as a platform for the direct exchange of information between users by enabling them to enter questions, cooperation wishes etc. ("Online Exchange"), to which any user of the network could respond.

On the one hand, setting up a network of this kind requires central servicing, coordination and attention; on the other, it has to have a built-in mechanism for respecting confidentiality. This can be achieved by requiring an appropriate and binding declaration from users, and by restrictive access controls. There are models for a virtual network of this kind in other European states, for instance at the Technology Partnership Centre of the Danish Technology Institute (DTI) in Aarhus. Experiences with these online-contact exchanges demonstrate a marked improvement in the exchange of information between enterprises, and facilitation in locating already existing solutions for problems which, without this forum, would have been far more difficult to find. In Bremen, too, various firms have articulated a demand for the interactive network, including firms who have been sceptical of the effectiveness of an exchange of information in this form. In the meantime, the online network "Bremen Business Net" (BBN) commenced operation. It is managed by the Bremen Innovation Agency (BIA) and accessible via the Internet.

The use of information and communication technology can certainly not be regarded a universal remedy for significantly improving an information exchange which has functioned badly up to now. Information deficits and contact barriers, on the industrial side as well as on the part of research, often have origins in personal behaviour which cannot be eradicated by technical services. Thus the BBN should not be misunderstood as a "patent recipe" for expanding and developing the contacts and cooperations formed by enterprises. It is a means of support intended to facilitate communication between the players in a regional innovation system. The more enterprises make use of the system, the greater the probability of assembling an appropriate range of solutions to problems. Whether the use of the network will remain at the level of a simple exchange of information, or whether, as a result of successful "mediations", closer cooperation will develop between users of the Online Exchange, giving rise to innovation networks, is a question that can only be answered after some time of operation.

6.   Conclusions

There are several different theoretical approaches that seek to describe and explain innovation-oriented cooperative relations between enterprises and their environment (for instance Håkansson's network theory, the GREMI concept of the innovative milieu, the theory of flexible specialization). Although these point to the relevance of networks for the innovative capability of enterprises, the ways in which cooperations arise - and particularly their structures and intensities - are the subject of controversy. This is clearly demonstrated by the assessment of Porter (1990), who sees the decisive impulse for the innovation and competitiveness of firms not in their cooperations, but in the interaction of four determinants:

- The extent to which factors relevant for the branch of the enterprise are present ("factor conditions");

- Demand conditions in the domestic market, and particularly the strength of international enterprises on the domestic market;

- Related and supporting industries resulting in mutual positive impacts through networks (clusters);

- Firm strategy, structure and rivalry.

Although clearly differing views of the same phenomenon - namely, the networking of firms in order to realize innovations - emerge from the various theories, there is consensus, according to the general status of knowledge in innovation theory, on the undisputed **necessity for firms to cooperate with various different partners in the innovation process**. This is made clear, for instance, by the "chain-linked" model of the innovation process put forward by Kline/Rosenberg (1986), according to which, in order to convert an original idea into a marketable product, horizontal and vertical networks and feedbacks occur not only between the different "phases" of the innovation process, but also between different institutions (e.g. enterprises, basic and applied research institutes, sales organizations) (Feldman 1994: 14ff.). This is not conceived as a static model; the links and feedbacks are determined by the problem to be solved and the configuration of partners.

Regarding their formation and structure, it can be concluded that the networks of enterprises are subject to a continuous process of change and learning and that they differ in character depending on the innovation project and the industrial environment (branch of industry) involved. As illustrated by the examples in this paper, they are associated with various initiation patterns and promotion approaches. Following Tödtling, the following **network differentiation factors** can be identified with regard to networking between high-tech enterprises (Tödtling 1994: 335ff.):

- Network structures vary between branches of industry. Contacts between firms in the sense of supplier-customer relationships and joint development projects dominate in the computer industry, whereas for biotechnology enterprises contact with academic and clinical (basic) research, investment companies and other finance partners is more important (cf. Reiß/Koschatzky 1997).

- Networks change in the course of the life cycle of an industrial branch. In the early growth phases, local and regional cooperative relationships generally dominate, whereas as the branch matures international market, supply and cooperation contacts are built up. However, there are also deviations from this pattern, as evidenced by the international contacts of many (new) biotechnology enterprises, for instance.

- Particularly in high-tech, networks can only develop in places where there are allocation advantages for certain branches (e.g. due to the presence of specialized suppliers of products and services). Thus these innovation networks do not usually develop spatially disperse, but tend to be regionally concentrated.

- From the viewpoint of regional policy, it is not only the presence of research and transfer institutions and other innovation services in a region which is important for the development of networks, but also their active cooperation and their receptiveness to the needs of the firms.

Networks represent important catalysts in the exploitation and strengthening of regional innovation potentials. Only by cooperation between enterprises and research and transfer institutions can synergies arise which have a favourable impact on the successful realization of innovation projects. However, since single regions cannot develop an industrial and R&D infrastructure that caters equally for the needs of all enterprises, network relationships should not - and must not - be confined to one region (interlinking between regional and national systems of innovation). The promotion of regional cooperation and networks must include the **bringing together of partners from different regions**. Only by integrating regional players into supraregional innovation networks can technology and development impulses be expected which can contribute to securing the national and international competitiveness of the enterprises within a region.

7. Bibliography

Aydalot, P. (Ed.) (1986): Milieux Innovateurs en Europe. Paris.

Camagni, R. (1991): Local "milieu", uncertainty and innovation networks: towards a new dynamic theory of economic space. In: Camagni, R. (Ed.): Innovation Networks. London, New York.

Coase, R.H. (1937): The Nature of the Firm. In: Economia (4), 386-405.

Feldman, M.P. (1994): The Geography of Innovation. Dordrecht, Boston, London.

Fritsch, M. (1992): Unternehmens-"Netzwerke" im Lichte der Institutionenökonomik. In: Boettcher, E./Herder-Dorneich, P./Schenk, K.-E./Schmidtchen, D. (Eds.): Jahrbuch für Neue Politische Ökonomie. Band 11: Ökonomische Systeme und ihre Dynamik. Tübingen, 89-102.

Gemünden, H.G. (1990): Innovationen in Geschäftsbeziehungen und Netzwerken. Working Paper, Institut für angewandte Betriebswirtschaftslehre und Unternehmensführung, Universität Karlsruhe. Karlsruhe.

Grabher, G. (1992): Entwicklung von Regionen-Netzwerke: Die Stärke schwacher Beziehungen. In: WZB-Mitteilungen 58, 3-7.

Grabher, G. (1993): The Weaknesses of Strong Ties: The Lock-in of Regional Development in the Ruhr Area. In: Grabher, E. (Ed.): The Embedded Firm - On the Socio-economics of Industrial Networks. London, New York.

Gundrum, U./Walter, G.H. (1995): Technologiepolitische Maßnahmen im Vergleich. Beitrag zur 3. Sitzung des Projektausschusses "Technologiepolitisches Konzept für die Steiermark. Manuscript. ISI: Karlsruhe.

Håkansson, H. (1982): International Marketing and Purchasing of Industrial Goods. Chichester.

Håkansson, H. (1987): Industrial Technological Development. Sydney et al.

Håkansson, H. (1989): Corporate Technological Behaviour, Co-operation and Networks. London, New York.

Håkansson, H./Johanson, J. (1984): Heterogenity in Industrial Markets and its Implications for Marketing. In: Hägg, J./Wiedersheim-Paul, F. (Eds.): Between Market and Hierarchy. Uppsala.

Håkansson, H./Johanson, J. (1993): The Network as a Governance Structure: Interfirm Cooperation beyond Markets and Hierarchies. In: Grabher, G. (Eds.): The Embedded Firm - On the Socio-economics of Industrial Networks. London, New York, 35-51.

Herden, R. (1992): Technologieorientierte Außenbeziehungen im betrieblichen Innovationsmanagement. Heidelberg.

Herrigel, G.B. (1993): Power and Redefinition of Industrial Districts: The Case of Baden-Württemberg. In: Grabher, G. (Ed.): The Embedded Firm - On the Socio-economics of Industrial Networks. London, New York, 227-251.

Imai, K./Baba, Y. (1989): Systemic Innovation and Cross-Border Networks. Paper presented at the OECD-Conference on Science, Technology and Economic Growth. Paris.

Kline, S.J./Rosenberg, N. (1986): An Overview of Innovation. In: Landau, R./Rosenberg, N. (Eds.): The Positive Sum Strategy. Washington, 275-305.

Koschatzky, K./Breiner, S./Gundrum, U. et al. (1993): Standortvoraussetzungen und Fördermaßnahmen für High-Tech-Unternehmen in der Region Rhein Main. Hauptstudie, Teil II zum Projekt High-Tech-Unternehmen in der Region Rhein Main im Auftrag des Umlandverbandes Frankfurt und der Wirtschaftsförderung Frankfurt GmbH. Frankfurt, Karlsruhe.

Koschatzky, K./Gundrum, U./Muller, E. (1995): Regionale Innovations- und Technologieförderung. Ansatzpunkte für die Nutzung regionaler Innovationspotentiale. Working Paper. ISI: Karlsruhe.

Lazerson, M. (1993): Factory or Putting-out? Knitting Networks in Modena. In: Grabher, G. (Ed.): The Embedded Firm - On the Socio-economics of Industrial Networks. London, New York, 203-226.

Maillat, D. (1992): La relation des entreprises innovatrices avec leur milieu. In: Maillat, D./Perrin, J.C. (Eds.): Entreprises Innovatrices et Développement Territorial. Neuchâtel.

Perrin, J.C. (1990): Organisation industrielle: la composante territoriale. In: Revue d'Economie Industrielle (51), 276-303.

Picot, A. (1981): Transaktionskostentheorie der Organisation. Hannover.

Porter, M.E. (1990): The Competitive Advantage of Nations. London.

Powell, W.W. (1990): Neither Market nor Hierarchy: Network Forms of Organization. In: Research in Organizational Behaviour (12), 295-336.

Quevit, M. (1991): Innovative environments and local/international linkages in enterprise strategy: a framework for analysis. In: Camagni, R. (Ed.): Innovation Networks. London, New York.

Reichwald, R./Möslein, K. (1995): Wertschöpfung und Produktivität von Dienstleistungen? - Innovationsstrategien für die Standortsicherung. In: Bullinger, H.-J. (Ed.): Dienstleistung der Zukunft. Märkte, Unternehmen und Infrastrukturen im Wandel. Wiesbaden, 324-376.

Reiß, T./Koschatzky, K. (1997): Biotechnologie: Unternehmen, Innovationen, Förderinstrumente. Heidelberg.

Rothwell, R. et al. (1972): SAPPHO Updated. In: Research Policy (3), 259-291.

Sabel, C.F./Herrigel, G./Deeg, R./Kazis, R. (1989): Economic prosperities compared: Baden-Württemberg and Massachusetts in the 1980s. In: Economy and Society (18), 374-405.

Schmidt, W. (1995): SPRINT-Programm Beratungsmaßnahme "Regionale Infrastrukturen und Strategien für Technologietransfer und Innovationsförderung" (RITTS). Projekt: RITTS 004, 1. Interim Report. Bremen.

Scott, A.J./Storper, M. (1990): Regional Development Reconsidered. Discussion Paper at the Workshop "Flexible Specialization in Europe". Zürich/Rüschlikon.

Sforzi, L. (1989): The geography of industrial districts in Italy. In: Goodman, E./Bramford, J. (Eds.): Small Firms and Industrial Districts in Italy. London.

Storper, M./Walker, R. (1989): The Capitalist Imperative. Territory, Technology, and Industrial Growth. New York, Oxford.

Storper, M. (1992): The Limits of Globalization: Technology Districts and International Trade. In: Economic Geography (68), 60-93.

Tödtling, F. (1994): Regional networks of high-technology firms - the case of the Greater Border area. In: Technovation (14), 323-343.

Willms, W./Färber, U./Hardt, U./Jung, H.-U. (1994): Konzept für eine regionale Infrastrukturpolitik im Raum der gemeinsamen Landesplanung Bremen/Niedersachsen. Band II: Wissenschaft und Forschung. In: Gemeinsame Landesplanung Bremen/Niedersachsen Schriftenreihe Nr. 5-94. Hannover, Bremen.

Wolff, H./Becher, G./Delpho, H. et al. (1994): FuE-Kooperationen von kleinen und mittleren Unternehmen. Heidelberg.

Technology and Incubator Centres as an Instrument of Regional Economic Promotion

Franz Pleschak

1. What are the Aims of Technology and Incubator Centres?

At present, there are about 180 technology and incubator centres (TICs) in Germany (cf. Figure 23). These accommodate nearly 4,000 innovative enterprises, representing a total of approximately 30,000 jobs. If associated enterprises, and those that have already left the centres, are included, this probably sums up to over 50,000 jobs. The centres house an average of 22 enterprises, with an average workforce of eight employees each. In the new Länder, more than 60 of these centres have been formed since 1990. The network of centres there is thus twice as dense as it is in the old Länder (cf. Schervier/Groß 1993).

The firms moving into the centres may be newly-founded technology-based firms, spin-off foundations from the R&D departments of larger firms, from public institutes or universities; technology-based service enterprises, contract research enterprises, qualification and further training institutions, or consulting firms.

The initiation and expansion of technology and incubator centres has the following aims:

- promoting firm foundations, particularly technology-based foundations, by providing an appropriate infrastructure, information and advisory services and creating favourable framework conditions for the development of these firms;

- rapid establishment of future-oriented technologies, by support of the transfer of knowledge, information and technology;

- initiation, strengthening and exploitation of synergies between science and industrial application, the networking of regional innovation potentials and the creation of national and international networks;

- development of the regional economy by the exploitation of skilled workforce structures; creation of new, attractive, innovative jobs, particularly in small enterprises; avoidance of "know-how-drains" caused by experienced, qualified people leaving the area, and support for setting up new businesses;

- advisory services for technology-based firms in the centres and elsewhere, e.g. help in drawing up business plans, dealing with legal and tax aspects, acquiring promotion funding, utilization of new technologies, and the divestment of enterprises.

Setting up technology and incubator centres is a part of national and regional policy for the promotion of industry and innovation. It is the task of technology and incubator centres, working in close coordination with universities and colleges of higher education, other research establishments and enterprises situated in the region, to create favourable conditions for the generation and marketing of innovations. The centres are intended to help focus the innovative potentials of a large regional catchment area and build up a regional profile in future-oriented fields of technology.

Due to their significance for the development of industry and innovation, technology and incubator centres have already been the object of several **empirical and theoretical studies** in Germany. The first results, relating to the old Länder, are in Sternberg (1988). There are also more recent results from empirical studies by Bauer/Hannig (1992); Steinkühler (1993); Pett (1994) and Sternberg et al. (1996). In particular, the Arbeitsgemeinschaft Deutscher Technologiezentren e.V. (association of German Technology Centres ADT) is able to provide a detailed current description of the existing centres in Germany and the enterprises working in them (cf. Baranowski/Groß 1996; Groß 1994; Innovationszentren 1993). Over the past few years, the Fraunhofer Institute for Systems and Innovation Research (ISI), Karlsruhe, in close collaboration with the economic geography department of the University of Hanover, has been concerned with questions relating to the initiation and expansion of technology and incubator centres in the new Länder (cf. Pleschak 1995; Tamásy 1995).

Figure 23: **Innovation centres in Germany**

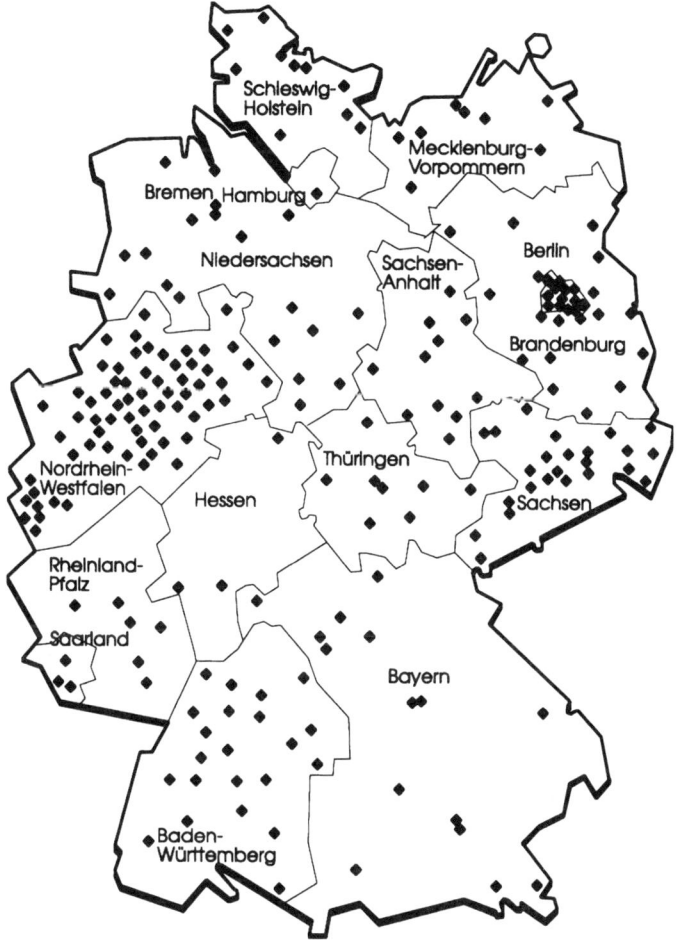

Source: Arbeitsgemeinschaft Deutscher Technologie- und Gründerzentren ADT e.V.

In the form of a model project, the Federal Ministry of Education, Science, Research and Technology (BMBF) has supported the start-up or expansion of 15 centres and the planning phase of a further 10 centres. ISI performed the scientific monitoring for this project. The statements in the present contribution are based on interviews with the managers of the supported technology and incubator centres, and on responses to a written questionnaire from 210 firms serviced by the centres, representing approximately 40 percent of firms renting premises in the centres at the time of the survey (Pleschak/Tamásy 1994). In the meantime, following these first examples, other centres supported by individual Länder or communities are in operation.

It is the purpose of this contribution to show how regional requirements must flow into the work of technology and incubator centres. This is important not only when elaborating planning concepts for new centres, but also in order to enhance the effectiveness of the centres in developing the regional economy. The latter aspect is actually more significant, as a high level of centre density has already been reached in the new Länder.

2. Main Aspects of Regional Surveys for Technology and Incubator Centres

In order to fulfil the aims pursued in setting up technology and incubator centres, these centres need to be regionally embedded. During **planning preparation**, the content and form of the concept for a technology and incubator centre (TIC) are already determined by the situation in the region, the types of initiating regional organizations, the regional conditions and the regional goals being pursued. Typically, at this preliminary planning stage the fundamental decision about setting up a centre is prepared in an initial step by a feasibility study, with only approximate analyses being carried out for many of the aspects needing investigation. Then, based on the fundamental decision, detailed concept and design studies are carried out. From these the business plan, the technical plan and the economic plan are elaborated (cf. Pleschak 1995).

The regional analyses shown in Table 23 are a prerequisite for the elaboration of planning concepts. Based on the results of analyses, the following decisions about setting up a centre can be reached:

Table 23: **Main aspects covered by regional surveys as a basis for decision about setting up technology and incubator centres**

- **Population structure**
  Number, age structure, migration tendencies, qualification structure

- **Potentials of labour market**
  Employment structure, specialization of workforce, qualification characteristics

- **Higher education and research landscape**
  Existing and planned universities and "Fachhochschulen" (colleges of higher professional training), research establishments, main technological emphases

- **Economic structure**
  Sectoral structure, development of trade and industry, employment by sectors, R&D intensity, demand for innovative services

- **Innovative Potentials**
  Source, orientation, innovative emphases, future emphases of technology, innovative infrastructure

- **Development of neighbouring regions**
  Existing and planned TICs, extent to which TICs are used; technological emphasis of regions, cooperation possibilities, foundations and development of enterprises, particularly those with innovation requirements

- **Economic promotion strategies**
  Promotion possibilities for founders, existing and planned trade centres, projects and initiatives to promote the economy, "marketing" of location

- **Chances for start-ups and allocation of technology-based firms**
  Potential of founders from universities, research establishments and enterprises, innovative infrastructure, ability and willingness to found firms, need for services provided by TIC

- **Regional interests**
  Initiatives by communities, chambers, higher education and research institutions, finance institutions and enterprises; attitudes of these institutions towards setting up a TIC, motivation structure, aims and individual interests, different location models

- aims in setting up the centre,

- formation of a business partnership by initiators,

- location and size of centre, choice of premises, building concept,

- technological profile,

- expansion stages, options for extension,

- financing needs and sources,

- time schedule,

- necessary networks,

- supply of services and consulting,

- structure of TIC management.

All regional organizations that have an influence on innovation development should work together in the **planning team** for a centre. The future managers of the centre should also be represented in the planning team, as the planning concept forms the basis for managing the start-up and expansion of the centre. During planning activities networks are already being formed with the community, the chamber of industry and commerce, with higher education and research establishments which will be important later on, both in consulting for firms and for regional economic development.

In interviews, centre managers mentioned the following factors as being particularly favourable when starting up and also for the **ongoing work of the centres** (Bachelier/ Pleschak 1994: 31):

- close cooperation with the chamber of industry and commerce and with local authorities,

- the presence of a qualified workforce in the region,

- intensive relations to a university, higher education or scientific institution in the region,

- competitive industrial potential in the region,

- a committed attitude on the part of initiators of the centre,

- a positive image of the enterprises and the centre.

The question whether a technology and incubator centre is significant for the development of regional technology and industry can be expressed in terms of the criteria given in Table 24.

Firms that have spent their start-up and development phase in a TIC can find the necessary opportunities for expansion in **technology parks**. In a technology park close to the TIC innovative firms, technology-based service suppliers, research and scientific establishments, branch offices of large enterprises, and firms previously accommodated in TICs can settle permanently. The closeness to the technology centre creates synergy effects and allocation advantages for these enterprises.

Table 24:   **Criteria for the regional effectiveness of technology and incubator centres**

- Creation of skilled jobs
- Improvement of allocation possibilities for technology-based firms, including founders from other regions
- Raising the technology level in the region, supporting regional economic structure, opening up new fields of technology
- Combating a migratory drain of R&D personnel
- Improving the quality of the location, enhancing the regional image
- Formation of a communicative "nodal point" for regional technology development, formation of regional innovative networks
- Contribution to the realization of regional development concepts and to the economic growth of the region
- Support of the transfer of personnel and technology
- Revitalization of disused industrial estates

Technology parks have a far greater influence on regional development than TICs. They give firms leaving the centres the possibility to expand their capacity, build up their manufacturing potential and market their innovative results. Young, expanding enterprises have room for manoeuvre when deciding on their future course of action. The possibility of settling permanently meets the firms' wish to own their own premises. Supply, service and cooperation relations grow up between firms within the park, and between the park and the centre. Enterprises located in the park generate a demand for technologies and innovative services. Research laboratories, university institutes, venture capital companies situated in the park, suppliers of finance services, marketing and service firms in the technology park can make their services available to innovative enterprises situated close by.

From a **regional viewpoint**, the expectations in setting-up a technology park are that the park should:

- contribute to renewing industrial structure,
- result in a new quality of the technological infrastructure,

- have a positive influence on the economic development of the region,

- result in the modernization and revitalization of obsolete industrial estates and former research institutions,

- limit public financial commitment to start-up financing, while mobilizing private capital to participate in the supporting companies.

In the **preparatory planning for a technology park** it is necessary to adopt a long-term planning horizon, since there will be lasting impacts on regional and urban development, on the quality of life in the region, the housing supply and the technological profile. For this reason, the future development of the region over the next ten to 15 years should be investigated. The park can help to improve the coherence of the region, if the whole complex - technology development - science - education/training - ecology - infrastructure - estates and business activities is considered as an integrated whole. This will only be possible if the preparatory planning process is used to coordinate the interests of all the participants and arrive at a standpoint of consensus. Personalities with a catalytic, integrative influence must continually force the pace of planning progress. Under no circumstances should a technology park ever be regarded as a "real estate business".

The image of a centre or park is decisively dependent on whether the firms accommodated there are economically successful, whether they spur development in the region and whether they have an impact on the innovation climate. Centre managers can influence the firm structure of TICs by their admission decisions. The firm structure should be such that the centre is shown to be a source of innovative progress. In cooperation with the Economic Geography Department of Hanover University, ISI investigated the extent of the technology base of the newly-founded firms that were tenants in TICs in the new Länder. The technological fields and areas of activity of the firms were used as criteria for evaluating their technology base (cf. Table 25), together with firm characteristics such as R&D-to-turnover intensity, high-tech rating of initial products and contacts with other R&D establishments (cf. Table 26).

Investigations show that the level of technology of the centres is set mainly by the firms founded as a result of the pilot scheme of the BMBF (Federal Ministry of Education, Science, Research and Technology) "Förderung technologieorientierter Unternehmensgründungen in den neuen Bundesländern" (Promotion of new technology-based firms in the new Länder, TOU-NBL). The proportion of these firms targeting foreign markets is also much higher than the average for all firms in the centres as a whole.

Table 25: **Shares of selected fields of technology and areas of activity of firms in technology and incubator centres (in %)**

Field of technology / area of activity	All firms accommodated in BMBF-supported TICs (n=210 enterprises)	Firms supported by TOU-NBL scheme in TICs (n=43 enterprises)
Environmental technology and analysis	19.0	7.0
Software tools/development	16.7	18.6
Management consulting, training	10.0	0.0
Building and building technology	10.0	2.3
Metrology	9.0	23.3
Medical technology	4.3	11.6
Optics, optoelectronics	1.9	9.3
Sensorics	1.9	4.7
Others	8.1	0.0

Data base: Hanover University, Economic Geography Dept.

Table 26: **Selected characteristics of the technology base of firms in technology and incubator centres (in %)**

Characteristics	All firms accommodated in BMBF-supported TICs (n=210 enterprises)	Firms supported by TOU-NBL scheme in TICs (n=43 enterprises)
Firms with an R&D-to-turnover intensity of over 8.5 %	55.9	97.5
Firms with completely or largely innovative products	29.0	62.8
Firms with frequent or occasional contacts to		
– universities	60.9	90.7
– non-university R&D institutions	40.5	58.2
– R&D departments of enterprises	40.0	59.1

Data base: Hanover University, Economic Geography Dept.

Technology and incubator centres, with their supply of services and their embedding into regional and supraregional networks, offer development chances to new technology-based firms. Table 27 clearly shows the **advantages,** but also the **drawbacks** that firms associate with renting premises in a TIC.

Of the **services and joint facilities** offered by the centres, firms make relatively infrequent use of

- preparation services for fairs,

- joint secretarial and reception facilities,

- the possibility of advertising in the foyer,

- accounting services and central data processing.

The **advisory services** most often used were the mediation of business contacts and the mediation of contacts to authorities and banks.

Table 27: **Advantages and drawbacks (characterized as "great" and "medium") perceived by firms (n=210) in technology and incubator centres**

Advantages	Share of firms in %
Availability of rented premises	86.6
Better publicity	55.7
Informal contacts with other enterprises	55.3
Reduction of fixed costs	52.9
Contacts with research establishments	40.0
Spatial flexibility	35.7
Drawbacks	**Share of firms in %**
No possibility of spatial expansion	24.7
Production possibilities poor or non-existent	8.4
Too many distractions from own work	8.1

Data base: Hanover University, Economic Geography Dept.

Table 28 shows the importance attached by firms to the TICs' offer of rented premises, services and consulting.

Table 28:   **Evaluation of the offer of technology and incubator centres by firms accommodated in TICs**

Component	Percentage of firms evaluating component as "very important" or "important "
Availability of rented premises	87.1
Services and joint facilities	58.1
Advisory services	19.6

Data base: Hanover University, Economic Geography Dept.

Innovations are only generated in a favourable **business atmosphere**. Responses showed that more than three-quarters of the firms felt the atmosphere was very good, or good, vis-à-vis other firms, and over 80 percent of firms considered that the atmosphere was very good, or good, vis-à-vis the management of the centres. More than two-thirds of the firms would rent premises in the same centre again. The internal problems that firms have are mostly connected with the market entry of their products or services. Financing problems rank second.

3.   Present Tasks for Increasing the Effectiveness of Technology and Incubator Centres

Since a high density of technology and incubator centres has now been reached in Germany, the central issue at present is not how to increase their number, but how to increase their effectiveness. The tasks necessary to achieve this are very closely linked with the aims of regional development. These tasks are:

• **Market and competitive orientation of the centres**

What sort of technological profile a centre should aim for, what innovative networks need to be formed and what regional development tasks should be accomplished, depends on the concrete market requirements and the region's possibilities. Although technology-based firms with high-tech products are active primarily on international markets, they still need competent regional partners for the trial and testing stages of new products and processes. Small innovative firms are involved in a cooperative division of labour with large enterprises. The profiles of the centres should be such as to create synergies. The centres have to be receptive towards spin-off foundations coming from regional research establishments and universities. Since many foundations have their origins particularly in these two types of institutions, the contacts listed in Table 26 are very important, and need to be consistently cultivated.

The fact that there is now such a high density of centres gives rise to a competition situation between different TICs. Since the economic viability of a centre is largely determined by the extent to which it is used, its image is not a matter of indifference. A centre's image is determined not only by offering enterprises premises at reasonable rents and saving them fixed costs, but also by the contacts it provides to scientific and communal organizations, by the existing networks and by the innovative working atmosphere.

A market and competition orientation today certainly also implies the need to investigate thoroughly whether or not it makes sense to set up a further technology and incubator centre in a region. This investigation involves an analysis of founder potentials, and also an analysis of potential founders' capabilities and their willingness to found firms.

- **Defining the field of business**

The tasks of a centre extend far beyond administering a property. Tenants expect an efficient, reasonably-priced supply of technical services as well as competent advisory facilities. Although the demand for such services depends very much on the stage of development of the tenant firms, their R&D intensity and their areas of activity, it is part of the business of running a TIC to offer services of this kind, also to external enterprises. When planning the form to be taken by services and consulting, centres should consistently be guided by the principle of customer orientation. Thus it cannot be very satisfactory from the viewpoint of centre management that - as stated in Table 28 - only just under 20 percent of enterprises assess this offer as being "very important" or "important".

It is also a task of management to give impulses for the development of cooperation, both within the centre and between resident and external firms. In their responses to the questionnaire mentioned above, three-quarters of the 210 enterprises state that they are cooperating with other firms in the centre. Almost all the firms exchange information and just under half of them are working on joint assignments either inside or outside the centre. One-sixth of firms have not yet developed any cooperations, but are interested in doing so. As shown in Table 27, the contacts to other enterprises, research establishments and universities mediated by the centres constitute an important advantage of TIC tenancy in the view of firms.

Technology and incubator centres demonstrate a high degree of competence in assessing the technological structures of a region. In cooperation with other organizations involved in regional technology policy, they are thus in a position to perform the following tasks reliably (Baranowski/Groß 1994: 40):

• project management in technology based joint projects,

• formulating policies for the promotion of technology,

• carrying out technology-oriented qualification measures,

• "marketing" the region as technology and innovation-oriented,

• supporting and monitoring technology initiatives,

• developing a technology park,

• initiating new forms of innovation financing,

• building up technology assessment and "early warning" systems.

• **Improving profitability**

The centres were created as instruments for promoting the regional economy. The setting-up of these centres is the object of promotion programmes of the Federal Government and the Länder (Koschatzky 1996). Particularly in the new Länder, the BMBF pilot scheme "Auf- und Ausbau von Gründer- und Technologiezentren" (Initiation and expansion of technology and incubator centres) has ensured that centres have been created very quickly, that experience from centres in the old Länder has flowed into the planning concepts, and that new firms have had a chance to locate there at all. Thus it is understandable that for the firms in TICs, availability of rented premises heads the list of advantages (cf. Table 27). In their start-up phase the centres receive public support. Subsequently, however, they are expected to be independent of ongoing financial assistance and to exercise freedom of action in their own management.

It is an essential prerequisite for the "entrepreneurial" management of technology and incubator centres that the tasks mentioned above in relation to market and competition orientation should have been successfully completed, and the fields of business defined. The fixed and variable costs of a centre should be covered by charges in line with market prices for consulting and services, by earnings from business activities carried out on behalf of the region, and by rentals and facility charges that are within the reach of new firms but are still cost- and demand-oriented. The most important factors influencing the costs are: the area capacity of the centre, the extent to which it is used, and the TIC management structure.

The administrative and service tasks of a centre can be facilitated by computer-aided support. Specially-developed software is available for this purpose, such as the programmes "TGZ-Manager" (cf. Mildner 1995) and "TZ-easy" (Lingen/Pagel 1994). Rational real estate management also helps to reduce administrative costs (cf. Platz 1993; Frutig/Reiblich 1995).

- **Organizing the technological information and communication infrastructure**

Asked for possible suggestions for improving the work of the centres, firms mentioned the following aspects most frequently as being "very important" or "important":

- stronger mediation of contacts to potential customers,
- stronger representation and publicity for the TIC and
- stronger representation and publicity for the individual firms in the centre.

Modern information and communication technology solutions can make a contribution to this aspect. Presenting enterprises and centres on networks like the Internet helps bringing partners together, supports firms in their innovative activities, makes information about research results more accessible, helps in project preparation. This type of presentation can be considered as being on a par with direct marketing activities (cf. Lohmann 1995; Tischler 1995). Efforts towards the networking of technology and incubator centres, such as those in North Rhine-Westphalia and Baden-Württemberg, for instance, aim in the same direction (cf. Ebers/Fricke 1995; Hoffmann 1995).

In Saxony there is a project to set up tele-services for enterprises and technology centres/agencies. According to Tischendorf/Uhlmann (1995) this might include the following types of information:

- Information on firms
  Firm profiles, presentations on products and services

- Information on technology and innovation
  Informative events, promotion programmes, R&D handbook, "technology mart", etc.

- Information on business know-how for development of the enterprise
  Buying, materials management, development, manufacturing, selling, marketing etc.

- Regional information
  Information from trade, industry and administration, connection with city/district information systems etc.

- "Supply and demand" mart

- Discussion forum
  Free inquiries and discussions

- Items of current information from firms, technology centres and agencies; coordination of the marketing of firms and centres with the marketing of cities and the region

- Online magazine
  Electronic access to available centre newsheets and transfer information circulars etc.

An information system of this kind is useful because it enables clients and cooperation partners to be found and selected faster and more accurately, allows users to expand their own marketing activities and creates synergy effects in the supply of products and processes. More time can be saved in the coordination of procedures.

4.   Summary

It is clear from this account that the initiation and development of technology and incubator centres is a component of regional technology and innovation concepts. As an element of regional innovation, centres are closely linked with its other elements (universities and higher education establishments, research institutes, technology transfer and advisory institutions, investment companies). The centres help to shape the infrastructure of a region, its innovation culture, its competitive and cooperative relations (Muller et al. 1994). In order to fully exploit the chances offered by the centres and achieve synergies between innovation actors in a region, there is a need for regional networks (cf. the previous chapter on the importance of innovation networks for small enterprises). Centres can enhance regional innovation potential, if they motivate existing firms to engage in innovation activities and stimulate the founding of new innovation-oriented enterprises. They strengthen the innovation infrastructure and supply innovation services. However, it is a prerequisite for all these beneficial effects that technology and incubator centres should be understood as an instrument for promoting the regional economy.

5.   Bibliography

Bachelier, R./Pleschak, F. (1994): Ergebnisse der Förderung des Auf- und Ausbaus von Technologie- und Gründerzentren in den neuen Bundesländern - ein Modellversuch des Bundesministeriums für Forschung und Technologie. In: Groß, B. (Ed.) Innovationszentren der 90er Jahre. Berlin.

Baranowski, G./Groß, B. (Eds.) (1996): Innovationszentren in Deutschland 1996/ 97. Berlin.

Baranowski, G./Groß, B. (Eds.) (1995): Innovationszentren in Deutschland 1994/ 95. Berlin.

Bauer, H./Hannig, U. (1992): Kritische Erfolgsfaktoren deutscher Technologiezentren. Studie der Wissenschaftlichen Hochschule für Unternehmensführung. Koblenz.

Ebers, H.-J./Fricke, J. (1995): Kommunikations- und Rechnernetze für das Netzwerk der Technologiezentren in Nordrhein-Westfalen. Paper at the 7th International ADT Annual Conference 1995. Berlin.

Frutig, D./Reiblich, D. (1995): Facility Management: Erfolgsfaktoren der Immobilien- und Anlagenbewirtschaftung. Stuttgart.

Groß, B. (Ed.) (1994): Innovationszentren der 90er Jahre. Berlin.

Hoffmann, H. (1995): Vorhaben zur Vernetzung der Technologiezentren in Baden-Württemberg. Paper at the 7th International ADT Annual Conference 1995. Berlin.

Innovationszentren (1993): Zehn Jahre Innovationszentren in Deutschland - Erfahrungen aus den ersten Technologie- und Gründerzentren. Berlin.

Koschatzky, K. (1996): Relationship between Universities, Enterprises and "Technologieparks": The Role of the Länder. In: Gaetano, G. de/Logue, H. (Eds.): RTD Potential in the Mezzogiorno of Italy: The Role of Science Parks in a European Perspective. European Commission: Luxembourg, 51-55.

Lingen, M./Pagel, H. (1994): Kosteneinsparung durch den Einsatz moderner Software in der Verwaltung von Technologiezentren. Paper at the ADT Annual Conference 1994. Aachen.

Lohmann, A. (1995): Elektronisches Infomations- und Präsentationsmarketing für Technologiezentren. Paper at the 7th International ADT Annual Conference 1995. Berlin.

Mildner, R. (1995): Informations- und Kommunikationstechnologien in Technologiezentren. Paper at the 7th International ADT Annual Conference 1995. Berlin.

Muller, E./Gundrum, U./Koschatzky, K. (1994): Horizontal Review of Regional Innovation Capabilities. Final Report. ISI: Karlsruhe.

Paschke, S. (1994): Untersuchungen zu den Zielen und der Wirksamkeit von Technologie- und Gründerzentren in der Region Eisenhüttenstadt/Guben. Diploma at the Faculty of Economics, TU Bergakademie Freiberg. Freiberg.

Pett, A. (1994): Technologie- und Gründerzentren: Empirische Analyse eines Instruments zur Schaffung hochwertiger Arbeitsplätze. Frankfurt am Main.

Platz, J. (1993): Immobilien-Management. Prüfkriterien zu Lage, Substanz, Rendite. 3rd edition. Wiesbaden.

Pleschak, F./Tamásy, Ch. (1994): Ergebnisse des Modellversuchs "Förderung des Auf- und Ausbaus von Technologie- und Gründerzentren in den neuen Bundesländern. 4. Analysebericht. ISI: Karlsruhe, Dresden.

Pleschak, F. (1995): Technologiezentren in den neuen Bundesländern. Heidelberg.

Schervier, W./Groß, B. (1995): Technologie- und Gründerzentren mit kreativer Struktur. Article in Frankfurter Zeitung, 23. November 1995.

Steinkühler, R.-H. (1993): Technologiezentren und Gründungserfolg technologieorientierter Unternehmen. Dissertation at the Economic and Social Science Faculty of the Christian-Albrechts-Universität zu Kiel. Kiel.

Sternberg, R. (1988): Technologie- und Gründerzentren als Instrument kommunaler Wirtschaftsförderung. Bewertung auf der Grundlage von Erhebungen in 31 Zentren und 177 Unternehmen. Dortmund.

Sternberg, R./Behrendt, H./Seeger, H./Tamásy, C. (1996): Bilanz eines Booms. Wirkungsanalyse von Technologie- und Gründerzentren in Deutschland. Dortmund.

Tamásy, Ch. (1995): Förderung innovativer Unternehmen durch Technologie- und Gründerzentren als Instrument der regionalwirtschaftlichen Entwicklung in Ostdeutschland. Dissertation at Hanover University. Hannover.

Tirschler, H.-D. (1995): Informationssysteme für KMU und ihre Nutzung im Technologiezentrum Eisleben. Paper at the 7th International ADT Annual Conference 1995. Berlin.

Tischendorf, D./Uhlemann, H.-J. (1995): Virtuelles Technologie- und Firmenzentrum (Südwest) Sachsen. Paper at the 7th International ADT Annual Conference 1995. Berlin.

Methods for Ascertaining Firms' Needs for Innovation Services

Emmanuel Muller, Uwe Gundrum, Knut Koschatzky

1. Introduction

In 1995, in connection with the European Commission's "European Innovation Monitoring System" (EIMS), a group of research and advisory institutes in Belgium, France, Great Britain and Germany compiled a set of handbooks for the support of regional actors in policy, industry and research in the development of regional innovation concepts and plans. These handbooks, which were intended for use in the RIS (Regional Innovation Strategies) and RITTS (Regional Innovation and Technology Transfer Strategies and Infrastructures) project promotions (cf. RIDER et al. 1996), provide methodological instruments and action recommendations. This set comprises **three parts:**

- analysis of the needs of small and medium-sized firms for innovation services,

- evaluation of the efficiency of the regional innovation infrastructure,

- analysis of technological and industrial development trends from the standpoint of the region.

The "Innovation Services and Regional Development" Department of the Fraunhofer Institute for Systems and Innovation Research (ISI) prepared the first handbook, dealing with the **systematic gathering of data on the needs of small and medium-sized firms (SMEs) for support** in their innovation activities. This involved the analysis of numerous empirical studies concerned with different methods of ascertaining the needs of SMEs for innovation services. The various individual approaches were summarized, forming the basis for the methodology described in the handbook (Muller et al. 1995). The systematic, continuous recording of the need of small and medium enterprises for support is **decisively important for the success of regional promotion of technology and innovation**; only in this way can promotion measures·be oriented towards the real needs of regional industry. In order to do this, the existing information resources of the region should be consistently exploited, judiciously combined and complemented. By creating an appropriate information system, in collaboration with the relevant actors in policy, industry, science and administration, the foundations are laid for the effective promotion of technology and innovation in the region. At the same time, the status of knowledge about regional innovation conditions is improved. This has the effect of enhancing the willingness of regional actors to engage in joint measures (regional enhanced sensitivity and activation). The needs of SMEs for support relate to all areas of industrial innovation, i.e. the development, production and market entry of new (technology-based) products and processes, as well as the exploitation of new production technologies and product elements in the activities of the enterprises. This involves the **whole spectrum of innovation services**: research and development (R&D), technology transfer, technology information and consulting, technical and business management training and further qualification, innovation financing and innovation management, start-up support, market research and marketing. These services are provided by public and private institutions, who are therefore also addressees for the action recommendations of the handbook, as are policy-makers at a regional level, whose influence can improve the general framework conditions for industrial innovation through public promotion programmes, the provision of a high-quality technical infrastructure (e.g. fast transport and telecommunication), industrial estates with an appropriate combination of production and service enterprises, good housing and leisure facilities, etc.

In the following contribution, the most important **methodological approaches to the systematic identification of the needs of enterprises** contained in the handbook are presented according to a unified schema and evaluated regarding their main characteristics, strengths and weaknesses. First, the general basis for research on the needs of firms in a region will be outlined and the main dimensions of the analysis described. **Four main methods of analysis** can be distinguished:

- collection and analysis of existing information,
- interviews with experts,
- personal interviews with enterprises,
- questionnaires addressed to enterprises.

Each of these methods has specific prerequisites and aims, and yields specific types of results. Finally the main characteristics, strengths and weaknesses of the approaches are presented in a comparative overview. The handbook is intended to assist regional actors in selecting, from among the analytical method for identifying the needs of firms, the one which is most suited to their aims and to the initial situation.

2.   Main Questions when Collecting Information on Needs

When **preparing an investigation** on the needs of firms in a region, the following questions need to be answered:

- What is the main objective of the investigation (e.g.: evaluation, expansion or optimization of the regional promotion system)?

- What are the main determinants of industrial innovation in the region (e.g.: regional structure of firm sizes and sectors, regional innovation services, qualification of workforce, communication infrastructure, etc.)?

- What resources are available for the survey (financial resources, but also personnel capacities and their level of qualification)?

- What types of partners can/should be involved (e.g.: chambers of industry and commerce, research institutes, associations, etc.)?

- What sources of data already exist, and can be used (regional data or other data that could be "regionalised")?

- Should the inquiry focus on a specific target group of firms (e.g. high-tech enterprises or firms in a specific sector), or should it include the regional economy as a whole?

- Is a static or dynamic analysis required (should the study have a short-term or a long-term perspective)?

- What form should the results take, and what are they to be used for (e.g.: recommendations for regional actors, or public studies to be used as the basis for a broader policy discussion)?

The answers to these questions provide a **research framework** on which the inquiry can be based; Figure 24 summarizes its basic elements.

Figure 24:   **Basic elements of an inquiry to identify the needs of firms for innovation support**

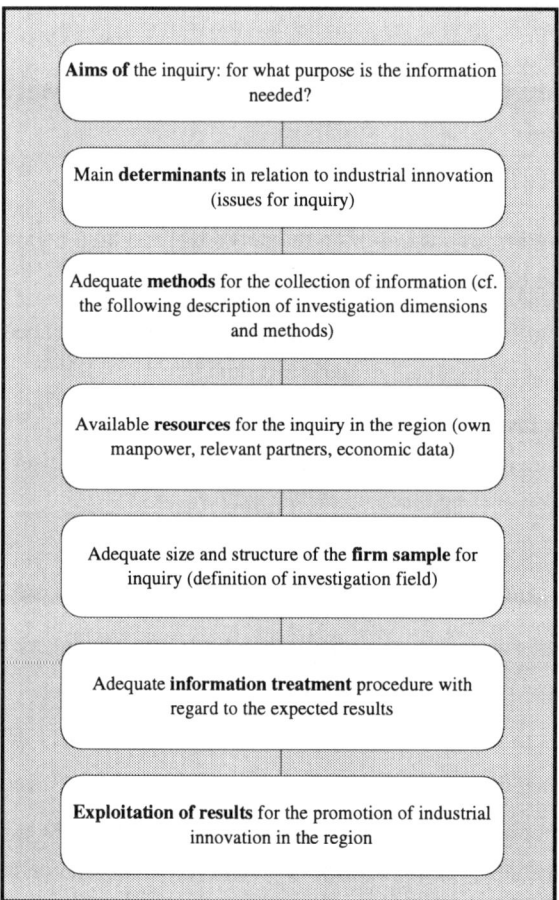

3.   Main Dimensions of the Analysis

The following **dimensions** are relevant when performing an analysis of firms' needs for regional innovation support:

- Firstly, the **spectrum of needs** to be analysed. The analysis may either be narrow and oriented only towards technical problems (technology transfer in the narrower sense), or it may also cover a wide range of needs relating to innovation activities (e.g. financing, marketing, general advisory services).

- The second dimension is the **degree of analytical differentiation**. An inquiry may be planned to yield highly differentiated results, for instance by firm sizes and sectors. However, it could also be envisaged to analyse the needs of enterprises at a purely regional level, without placing any particular emphasis on different firm characteristics or concentrating on different types of needs.

- The third dimension of analysis relates to the **degree of expression of the needs of firms**. Studies usually concentrate on the explicit (i.e. expressed) need for support. However, a situation may also arise in which firms, although they are unable to formulate their innovation-related problems, still have a latent need for support which they themselves are not able to recognize clearly.

A regional inquiry should **take account of** the following **aspects**:

- it should gather information on a broad spectrum of needs which does not only include technical issues;

- it should aim to provide results which are differentiated with regard to regional firm structure;

- it should include an analysis of enterprises' latent needs for support.

4. Methodological Approaches to the Analysis of Needs

This chapter summarizes the **methods most frequently used** to identify the needs of small and medium-sized firms for support:

- the collection and analysis of existing information,

- interviews with experts,

- interviews with representatives of firms,

- written questionnaires addressed to firms.

As well as these methods, other approaches are also possible (e.g. "technology audits" of firms; observation of enterprises over a period of time by specialist institutions; expert workshops, etc.); here, however, only those methods are presented which can be applied without substantial external assistance. The individual methods are not mutually exclusive; the parallel use of several methods can help to enlarge the information base.

The four methods presented here can be classified into **two groups** (cf. Figure 25):

- Indirect collection of information on the needs of firms, based on already existing knowledge sources;

- direct collection of information on the needs of firms, based on inquiries to the firms themselves.

In the following section a brief outline of each method is given. This is followed by a description of the most important methodological steps involved, a discussion of the type of results to be expected from each method, and the resources it requires.

4.1   Collection and Analysis of Available Data

The collection and analysis of information from various existing sources at a regional or national level (e.g. data on the demand for financial support for innovation projects, or for R&D personnel) can be used to identify the needs of firms. This method is based on the **collection and classification of available information**. It can be complemented by continuous information gathering (or by new inquiries designed to fill the information gaps that have been found).

Figure 25:   Indirect and direct identification of the needs of firms

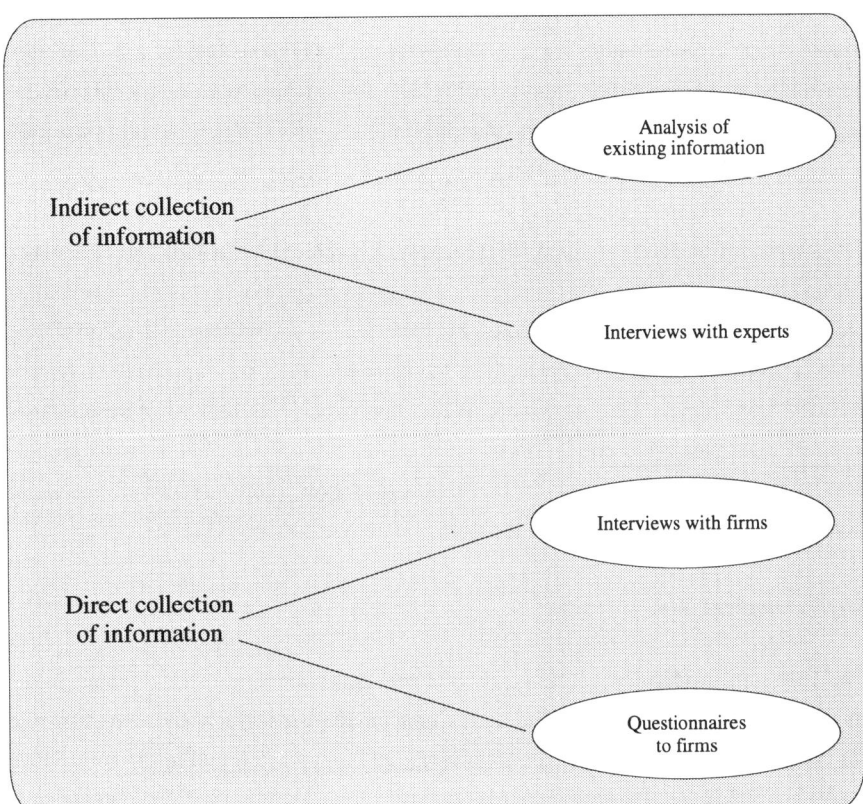

General description

One of the main advantages of this method is that its **costs** are **low**, since it is based on exploitation of the results of already existing studies and questionnaires and the analysis of promotion projects, thus creating **synergy effects**. Another positive effect is the enhanced motivation and awareness of regional institutions and authorities induced by the cooperation that has been necessary in order to enable these information collections to be made. This approach does not allow a comprehensive analysis of different types of needs; it is dependent on previous studies and already existing models. Moreover, on grounds of confidentiality it is sometimes difficult, or even impossible, to obtain certain items of information.

This method can only be expected to provide a part of the information relating to the needs of firms. However, due to the positive impacts of the exchange of information and better resulting adaptation of the innovation support system to the needs of firms, it should not be underestimated. As well as the existence of a data base in usable form, a coordination team is necessary, to be responsible for the collection and analysis of information (and possibly also for carrying out complementary inquiries). This team should consist of representatives of various regional innovation actors, in order to create more effective cooperation within the region.

The **spectrum of needs** covered by this method is relatively **limited** and is dependent on the data available. The method concentrates mainly on expressed demand; if the data have been initially collected by experienced technology consultants who have helped the firms to express their latent needs, then this approach may also identify latent needs. One drawback of this approach is its lack of representativeness. In most cases, the collection of information can only include a small proportion of the population of firms in the region (however, there may be strong variations in this aspect, depending on the type of data recorded).

Methodological steps

The first step in identifying the needs of firms and collecting other relevant information consists in mobilizing the actors concerned (particularly those responsible for supplying support services for technology and innovation) and in the continuous analysis of various information sources in the region (and possibly also outside it).

The second important step is to form a team (which may also include external advisors) to take on the collection, organization and analysis of information.

The **collection and analysis of available information** should be organized as follows:

- inventory,
- classification,
- evaluation.

Evaluation of the utilisable information can take place in a **workshop** with the regional innovation actors (enterprises and providers of innovation services). If the available information is not sufficient, new investigations can be initiated to complement the existing data. One strategy which should be considered is to build up a data base in which the various types of information are stored in a more or less standardized form. Decisive for the success of this approach is the **continuous analysis** of information, synthesising the most important results so that recommendations for action can be derived from them.

Another possible strategy is to install a **permanent regional observation unit** with the task of collecting and analysing information on the needs of firms. The analysis of existing information, as described above, could be a first step in this direction.

4.2   Interviews with Experts

Interviews with experts have the purpose of identifying the indirect needs of firms by the **exploitation of expert knowledge and previous studies.**

General description

It is a prerequisite for this approach that valid information on the region exists, and that there is a sufficient number of experts. If these conditions are fulfilled, this method allows the **comprehensive gathering of experiences, opinions and the results of analyses.** Its comprehensiveness is enhanced by the possibilities of an expert workshop with direct communication. The approach is also characterized by a high degree of flexibility (in the issues addressed and the people involved) and by relatively low costs.

As a result, a **complete picture or corresponding "vision" of the needs of firms** within the region can be expected, based on the experiences and knowledge of the experts. There should be a possibility to bring in experts from outside the region, provided they represent a reliable source of information for several regions. There is generally a need for specialists to analyse the information gathered and document the results systematically. If there is a broad-based selection of experts, a broad spectrum of needs will also be covered. Depending on the information base, the expert knowledge and the range of topics selected, this procedure makes it possible to identify expressed demand and - to a limited extent - latent needs.

Figure 26: **Collection and analysis of available data**

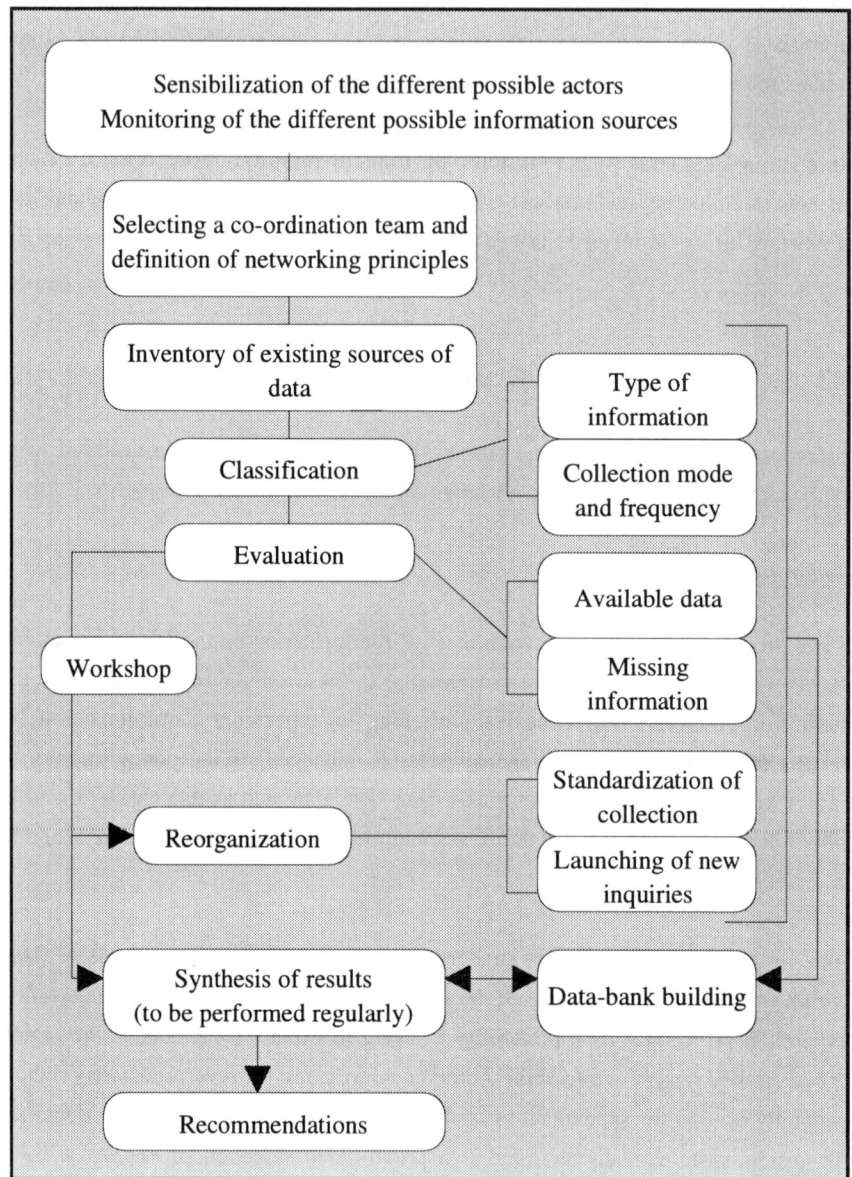

Methodological steps

The preparatory tasks consist in forming a research team, selecting the experts to be interviewed and defining the procedures for collecting information. These are the basic elements that determine the success of the action. The selection of experts should include the following:

- regional administrators,
- representatives from industry,
- scientists,
- consultants/advisors.

The expert interviews should allow a free and open **exchange of information** on the issues discussed. The last point is the **summarizing and analysis of information**. This should express the basic ideas and opinions of the experts interviewed, but should also take controversial positions into account for inclusion in later consensus-forming.

Figure 27:   Organization of interviews with experts

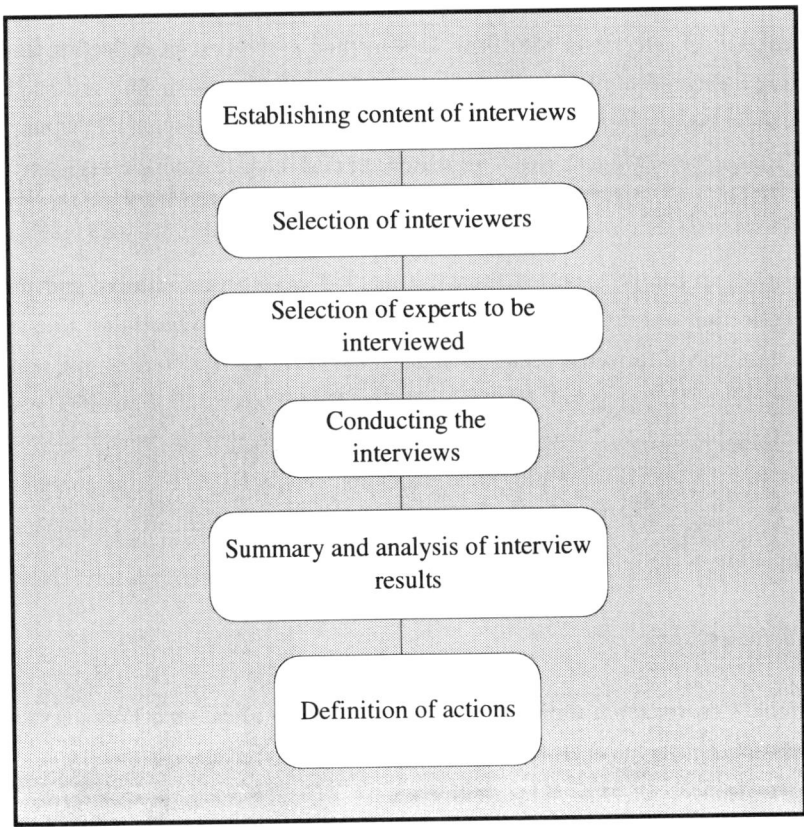

4.3   Interviews with Firms

In this approach, the needs of firms can be identified by **personal interviews with managers or specialists from the firms,** either

- conducted on the basis of structured questions ("semi-directive" interviewing), or
- without standard questions (open interviewing).

General description

Personal interviews with representatives from enterprises allow a relatively **complete and open collection** of information about needs. The quality of the results depends on the interviewer's possibilities for gathering detailed and accurate information and discussing it with representatives of the firms, in order to be able to interpret the firms' needs. Standardized question forms simplify the interpretation of the information gathered.

The weakness of interviews with firms as a method is the **cost of collecting and analysing information** (due to the fact that this is very time-consuming and qualified personnel is required for collection and interpretation). Cutting down the number of firms selected for interviews for reasons of cost leads to a non-representative survey, and thus to results with a limited relevance.

The results that can be expected from this method are a **representative and detailed collection** of information on the needs of enterprises, provided that a representative sample of firms has been selected and the range and structure of questions are appropriate. This approach generally requires the involvement of qualified personnel. External advisors or representatives of regional institutions should also be brought into the interviewing procedure. Attention should be paid to the composition of the sample, which should depend on the sectoral and firm size structure of the population of firms in the region.

Methodological steps

In an initial step, the **main topics** of the interview have to be defined. The interview procedure may have one of **two basic forms:**

- open interviews (general discussion without pre-structured questions),
- structured interviews based on a standardized pattern.

The next step is to form a team responsible for the collection and analysis of information. If the necessary know-how is not available within the institution running the project, it is advisable to bring in external consultants. The selection of firms to be interviewed is decisive for the success of the procedure. Selection should have the aim of compiling a **representative sample** in consideration of regional firm size and sectoral structure.

The **size of the sample** is decisive for its representativeness. It is also important to find a **competent interview partner** within the firm who has a good overall view of the firm's development and its present situation: the most suitable partner is the general manager or owner of the enterprise. It is necessary to **prepare the interview partner** by providing preliminary information on the main questions.

The interview should not be conducted separately from the **interpretation of results**, which should involve the same persons. In this way there is interaction with the interviewed firms during the information analysis phase; thus enabling a more complete assessment to be made of the needs of the firm ("feedback loop").

4.4　　Written Questionnaires to Firms

In this approach, data on the needs of firms in a region are collected by means of a detailed **written questionnaire**. Statistical methods are then used to analyse the information gathered.

General description

Written questionnaires can achieve a **high degree of representativeness** by the detailed recording of data on firms in the region (differentiated according to firm sizes and sectors). Representativeness depends on the response rate to the questionnaire. Standardization of questions and pre-arranged answer categories facilitate analysis of the information collected.

Figure 28:   **Organization of interviews with enterprises**

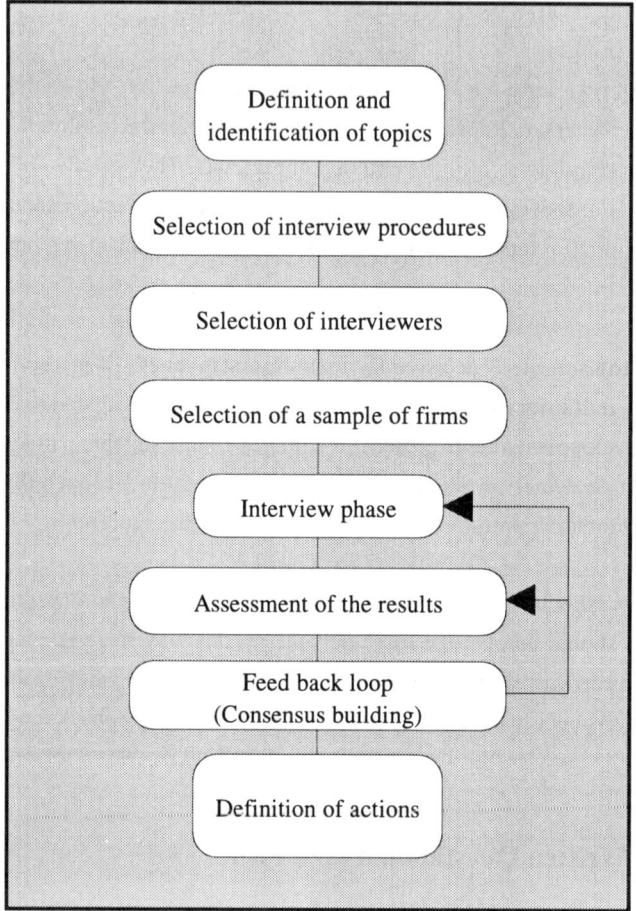

The method's main weakness lies in the **lack of depth of the information gathered.** Another problem arises from the need to aggregate (with the result that part of the information is lost). Because of the large number of responses expected, analysis should be performed using descriptive and multivariate statistical methods. With written questionnaires there is a danger that the **response rate** may be **too low** (particularly in areas which have been "over-researched", or if the questionnaires are too complex in design). Questions must be formulated in such a way that there is a unified understanding of their content among the firms addressed.

Written questionnaires allow the **identification of the explicit needs of firms.** They can help to discover the main deficits in the regional innovation support system. It is necessary to form a coordination team (including external consultants), to be responsible for:

- preparing the questionnaire,
- encoding the questions,
- performing statistical analysis,
- analysing the results and formulating the conclusions.

The spectrum of needs that can be recorded by means of a written questionnaire is necessarily narrow (because of the necessity of focusing on some specific issues). Although data can only be collected on explicit needs, it is possible to achieve a high degree of representativeness due to the great number of firms that can be reached in this way.

Methodological steps

Information can be collected not only from **written** questionnaires, but also from **telephone** interviews or a combination of **written questionnaires and telephone** interviews. Telephone interviews can yield a higher "response rate", but they are more time-consuming. There is also a qualitative difference between the answers obtained by the two methods: in telephone interviews there is a possibility of explaining the questions and examining interrelations. The basic principles of questionnaires and telephone interviews are similar. A survey should begin with the definition of the target groups and the content of the questions. The next step is designing the questionnaire, which is the standard vehicle for collecting information. A questionnaire should not be too long (not more than 4 to 6 pages) and should be as comprehensive as possible.

Particular attention should be paid to the layout, in order to make the questionnaire simple to answer and improve the response rates. It is a good idea to test it out on a small number of firms (pre-testing) so that corrections and improvements can be made. Following the pre-test, the questionnaire can be sent out to the firms, preferably accompanied by a covering letter of explanation giving the name of a contact partner who can deal with queries.

After the return of the questionnaires it is important to ensure representativeness of the results, if necessary by organizing follow-up actions ("reminders") in order to arrive at a balanced sample of responding firms and achieve an adequate response rate (in consideration of the region's firm size and sectoral structure). Analysis of the information collected begins with the encoding of the returned questionnaires. Subsequently statistical processing (mainly descriptive statistical and factor analyses, possibly econometric regression) enables a set of results to be drawn up which is interpreted and assessed by experts (possibly in the form of a workshop). Finally, the results of the questionnaire should be summarized and presented in a suitable form (at an informative event, or in a report); this final step should also include the presentation of concrete recommendations for regional actors.

Figure 29:   Organizing a questionnaire addressed to enterprises

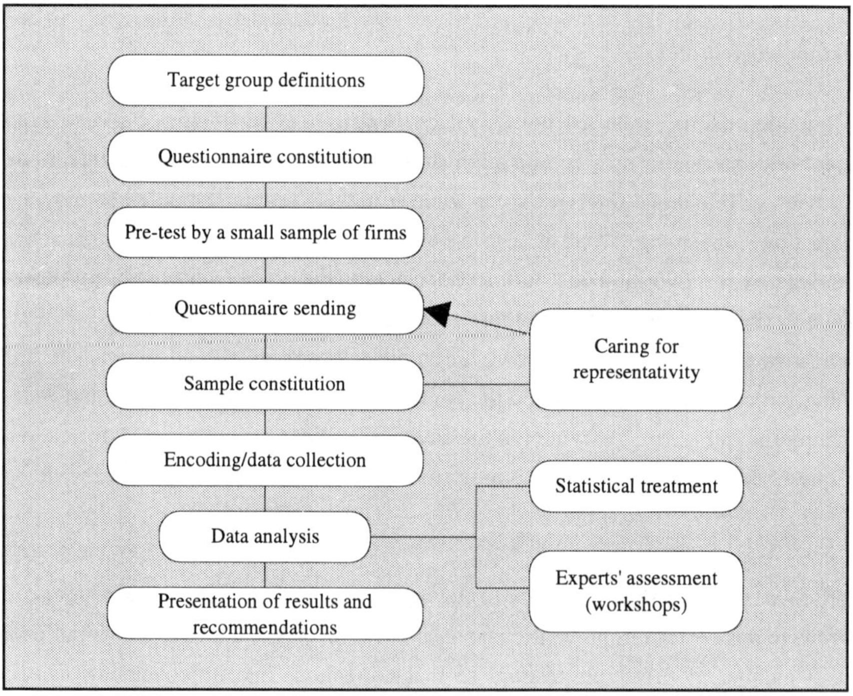

5. Evaluative Synopsis

The following synopsis shows the main characteristics, strengths and weaknesses of the methods described for identifying the needs of enterprises for regional innovation support:

	Analysis of existing information	Interviews with experts	Interviews with firms	Questionnaire addressed to firms
Strengths	Synergy effects	Consensus formation and mobilization	Gathering of public information	Standardization of questions (facilitating analysis)
Weaknesses	Dependent on existing structures	Results difficult to evaluate (subjective statements)	Small number of firms (low representativeness)	In-depth analysis not possible
"Input" required (relative to "output")	+	+	++	+
Demand spectrum	--	+	++	-/+
Identification of latent demand	--	+	+	-/+
Representativeness	--	-	+	++

++ good   + fairly good   - not very good   -- poor

The approaches described can be generally recommended for the analysis of firms' needs for regional innovation support. The synopsis of the main strengths and weaknesses of the various approaches, their costs, the demand aspects they cover, and their representativeness, is **intended to help regional actors select the appropriate analytical method** in consideration of the specific conditions in the region. Each region has to decide which approach is suitable in view of its initial situation, available resources and previous experiences. All in all, these **general recommendations** can be made:

- Several different methods should be combined, in order to obtain information which is as detailed and comprehensive as possible;

- Particular attention should be paid to the methodological aspects of information collection and analysis;

- In the collecting and analysis of information, particular consideration should be given to the qualification of the project workers.

6.   Bibliography

Muller, E./Gundrum, U./Koschatzky, K. (1995): Methodology in Design, Construction and Operation of Regional Technology Frameworks. Needs analysis of the innovation and technology support requirements of firms within a region. Final Report to the European Commission, DG XIII-D. ISI: Karlsruhe.

RIDER (Coordination) (1996): Methodology in Design, Construction and Operation of Regional Technology Frameworks. Final Report. Louvain-La-Neuve.

The Adaptation of German Experiences to Building Up Innovation Networks in Central and Eastern Europe

Günter H. Walter, Ulrike Broß

In modern countries, industrial innovations ensure continuous adaptation to changing societal conditions, as well as securing, economic performance, national income and political weight in foreign affairs. For an innovative national economy, it is important that all a country's available resources should first be mobilized and interlinked, i.e. that science and industry in particular should combine their efforts in the realization of industrial innovations. In this respect, technology policy measures can support enterprises and research institutions in initiating and extending an innovative network and thus bring resources together. Since this is difficult in Central and Eastern Europe under the present conditions of radical change and re-orientation, Germany is providing "help towards self-help".

1. Central and Eastern Europe: the Path Towards International Competitiveness

With the cessation of the previous socialist order, the Central and Eastern European countries were directly confronted with the task of fundamentally transforming their political and societal systems. This systemic transformation is historically unique in its radicality. Against a background of political instability, a positive development of the economy acquires particular importance. The economic systems of these countries are now re-orienting, away from a more or less centrally planned economy and from embedding in the socialist state system towards a free market economy; they are opening up to the global economy. At present, most of them are only able to gain a modest foothold in the world market.

In the time immediately following the changeover, national and international initiatives were simultaneously faced - as they still are today - with a multiplicity of tasks in conditions of extreme shortage of resources. No clear strategies or policy recommendations could be derived from existing political and economic models. However, in view of the international competition of locations for scarce resources, an overall political concept seems to be a prerequisite successful economic development. An integrated plan to re-structure the whole economic order is more relevant than the use of individual policy instruments. Three influences are shaping the future order: firstly, parts of the old system survive, with their actors and institutions; secondly, economic policy models are being adopted which are mainly western in character and, thirdly, endogenous potentials are free to develop in a way which was not possible under the imposed former order. The transformation of an economy requires action at very different policy levels. One important starting-point for a country's innovative power and its economic success in the long term is innovation and technology policy.

Technology and innovation policy includes all public measures which are oriented towards converting technical inventions into industrial applications and which support the diffusion of product and process innovations (Meyer-Krahmer/Kuntze 1992). Instruments of public technology policy include measures such as the institutional support of research institutions, financial incentives for industrial innovation projects, and the initiation and expansion of an innovation infrastructure in terms of consulting, technology transfer and innovation financing.

In the following contribution an attempt is made to pick out, from the accumulated knowledge and concepts relating to success determinants in innovation and technology policy those that can appropriately be adapted for use in the process of modernization now taking place in Central and Eastern European countries. Technology and innovation policy in these countries clearly has to pursue strategies of modernization which help to open them up internationally, which mobilize their endogenous resources and form the basis for an integrated policy concept. Western countries can only provide very limited assistance for this transition process, in the form of "help towards self-help". But also this approach always has to take account of the differing initial situations in the individual Central and Eastern European countries and the differing paths they have adopted in the process of reform.

2.   A New Paradigm of Technology Policy

By opening up their economic systems, the countries of Central and Eastern Europe have joined the international competition. After several decades of political and economic encapsulation from the western world, a consensus may be said to exist with regard to the necessity for their integration into the global economy and into the international division of labour. This integration can best be realized by a "free market" type of economic system which derives its impetus from individual, decentral initiatives. The realization of existing competitive potentials is complicated by the state of the markets, which are not yet fully able to function, and by the existence of certain types of market failure. This situation also implies the need for a public technology and innovation policy. However, before going on to consider what instruments should be used in innovation and technology policy, and to what extent the state should intervene, it is necessary to consider what modernization strategy should be adopted in view of national circumstances.

Integration into the international division of labour can occur in different ways, as various economic theories suggest. The competitiveness of a country is based on the endowment of certain production factors, or access to them at low cost. Two examples of this are the predominating export of cheap raw materials currently observed in Russia, and low-cost manufacturing on a sub-contracting basis (extended workbench) in the low-wage countries. However, as a rule it is only possible to emulate with the leading industrialized nations or approach their standard of living by producing innovative products that are mature for the market. Unlike the export of raw materials, in which profits are strongly dependent on price fluctuations on the world market, the high domestic value-added resulting from the manufacturing of high-quality products, and the national income associated with it, are attractive from the viewpoint of any country. Thus in order to be suitable for countries in Central and Eastern Europe, a strategy must take account of insights into the creation and appropriate use of human capital, a factor which is significant for innovative economies since it is decisive in high-quality production.

Recent empirical and theoretical research offers starting-points for the derivation of strategies and policy recommendations. One basic element here is regional and national innovation networks. On the one hand, the development of modern technologies is characterized by a growing interdependence and complementarity of different areas in industry and society. Studies confirm the relevance of these factors for the economic success of a location. Well known examples in Germany are the Ruhr Valley area, East Bavaria and Baden-Württemberg. In regions which are economically highly-developed, networking between the different actors is very strong (cf. for instance Herden 1992).

Management research has thoroughly investigated the importance of strategic network relations, particularly between different steps of the value-added chain (Sydow 1992). Networks serve the purpose of interlinking actors in production, services and research in such a way that their comparative strengths are exploited to the full and developed further. Innovation networks are able to activate, coordinate and combine the resources which support the technological competitiveness of regions and countries.

Economic theory advances a valid explanation for this phenomenon in the network theory (Håkansson 1989). Generally speaking, in these models complementary - and therefore resource-saving - learning processes are initiated between the actors in the economic process (cf. also the chapter on innovation networks in this reader). Firstly, "learning by doing" and "learning by using" take place between suppliers, producers and customers in their business relations (Kline/Rosenberg 1986). But networks are not only characterized by performance-related business relations. In the course of repeated interchanges the integrity of the partners is recognized; a relationship of trust is built up, and stable informal or personal relationships also grow up between the partners (Walter 1992). Through the cooperation with the network partners, knowledge is accumulated which forms the basis for future competitive advantages. Externalities in an alliance of this kind extend beyond the reduction of transaction costs: dynamic economies of scale in terms of learning and complementary investments enhance the productivity achieved with limited resources. Moreover, the reduction of uncertainties by institutionalization of the exchange relationships, and the distribution of risks among several partners in case of failure, are very important aspects, particularly in industrial innovation activities. These are the factors that make networks so successful and have caused them to be an object for economic and regionally-oriented research for several years now.

The results of various studies indicate that the parameters of firm size, industrial sector and technology orientation give rise to different patterns and intensities of cooperation (Koschatzky et al. 1993). Innovation networks are based on specific national, regional or local development patterns in the sense of different "best practices" in technogenesis and technology use; the starting-point for these is formed by specific innovation potentials, such as accumulated knowledge in certain technologies. A synergetic innovation network arises e.g. through the alliance of regional actors in research, production and services, aiming at the optimal exploitation of existing resources for growth (Koschatzky et al. 1993). Overlapping, flexible network relations have proved to be effective in coping with structural change due to changes in the frame conditions for competition (Herden 1992).

From the viewpoint of innovation economics, national and regional networks can be used for the economic development of countries in Central and Eastern Europe, to systematically exploit existing development potentials by their conversion into application-oriented knowledge and the rapid diffusion of new technologies (Koschatzky 1995), thus also providing a good basis for participation in international networks. The innovation networks that are of particular interest for innovation and technology policy arise through cooperation between suppliers and users of technological knowledge: these may be relations between enterprises and suppliers of innovation services (e.g. universities and research establishments).

There are specific German experiences in network research. The innovation networks of the region of Baden-Württemberg (Herden 1992) are characterized by strong specialization of the firms at a high level of technology, by the heterogeneity of the industrial sectors they belong to and by the international orientation of regional industry. The innovation networks show a remarkably balanced combination of production and service firms and regional research establishments. Spatial proximity of the network participants has proved to be a particularly important factor, especially for small enterprises. Information technologies and the technical transport links between the various locations also play a very important role. The long term orientation of the technology and innovation policy of the "Land" of Baden-Württemberg has also contributed to the formation of networks. Thus the region gives comprehensive support for technology transfer and technology consulting via a non-profit foundation (the Steinbeis-Foundation for Economic Promotion), joint research and contract research projects, and contract research institutes (Landtag von Baden-Württemberg 1993). An "Innovationsbeirat" (innovation advisory council) consisting of high-ranking representatives from industry, science and society supports the Land Government in the integrated planning of this policy and in the ongoing development of its perspectives.

The innovative networks in Baden-Württemberg have also served as a model for other regions and countries, mainly due to the great variety of approaches adopted in Baden-Württemberg's networks, the different types of relations between the actors, and the experiences gained from them. Interested countries and regions have generally adopted the elements of networking that are most appropriate for their situation and have adapted them to their societal conditions, which in some cases differ very markedly from the highly industrialized region of Baden-Württemberg (as for instance in structurally weak regions in the new Länder or in Third World countries). Often the Steinbeis Foundation only mediates the first direct contacts between research and industry, leaving the further planning of the cooperation to the partners themselves. In this way, networking relations arise which - albeit initiated by the state - are autonomously organized according to free market principles.

The network concept can provide starting-points for the application of a specifically-oriented innovation and technology policy: firstly, the mobilization and judicious complementation of regional resources for the development and application of new technologies; secondly, the coordination and interlinking of these resources within innovation networks involving all the relevant actors in industry, science and policy; thirdly, the integration of these regional networks into the national and international development and production of technology, by the creation of active interfaces and the support of supraregional and international cooperation.

3. Utilisation of Experience for the Creation of Networks in Central and Eastern Europe

The network approach is a model which appears to be suitable for technology and innovation policy in Central and Eastern European countries.[1] Above all, it specifically induces coordination and cooperation (the interlinking of actors and potentials), giving rise to the complementary exploitation of resources and generating synergy effects. The theoretical network model includes structural factors and allows an assessment to be made of the contribution that networks can make to innovation-based positive economic development. In view of the numerous and complex interactions involved in economic transition, network theory represents an operational starting-point for the derivation of policy recommendations designed to deal with present-day problems. Central and Eastern European countries are now in a situation where their industry has to "catch up" under severely restricted conditions, caused by extreme shortage of resources associated with a high degree of uncertainty. German and international experience in networking research has shown that success in catching up can be achieved particularly through the activation of endogenous innovation potentials.

The following section commences with a brief description of societal framework conditions in Central and Eastern European countries. Following and based on this, networks are then specifically described and examples are given of possibilities for their implementation in Central and Eastern Europe.

3.1 The Present Situation in Central and Eastern Europe

Under the socialist economic order, research and industry in the countries of Central and Eastern Europe were characterized by a centrally-steered, "top-down" type of organization which was predominantly state-controlled. This situation led to vertical structures in science and industry, with very few relations of horizontal interchange.

---

1 Central and Eastern European countries are states situated to the west of the previous Soviet Union, plus the neighbouring countries of the Commonwealth of Independent States of the former Soviet Union (CIS) (Belarus, the Ukraine) and the Baltic States (Estonia, Lithuania and Latvia).

Exchange within networks brings the interaction between the various actors in the national economy to the fore. Thus what is required is a policy strategy emphasizing the free development of individual actors according to the "bottom-up" principle, in place of the former centralistic policies. In the foreground is the promotion of efficient, innovative cooperations in the form of "horizontal" relations between research and industry. In strategic terms, this implies that policy should aim to support interactive relations with a view to the formation of networks, and should itself become active in the creation of new networks.

In Central and Eastern European countries, innovation and technology policy measures that are intended to support networking activities are sometimes still influenced by the surviving remains of inherited political and legislative framework conditions. Another problem has arisen due to development in the first few years of the transitional phase when, due to the shortage of resources, drastic cutbacks in funding took place, with the result that today many previously-existing potentials in science and industry are on the point of collapse. Neither strategic plans nor financial means were available for their consolidation during the system changeover.

Industry and science

Industrial structure in the countries of Central and Eastern Europe, with its large-scale business units (e.g. combines), is because of the size scale disproportionate to the small domestic markets. Capital-intensive production was previously based principally on tayloristic mass production and economies of scale; only exceptionally were products offered which satisfied international standards. Production plants today are mostly obsolete, new technologies are scarcely used due to the limited financial resources. In all Central and Eastern European countries, the macro-economic situation is marked by drastic declines in production in the first few years of transition, due to the disappearance of the trade relations formed under the socialist regime.

Today, as well as the large enterprises, there are often also a wide variety of small business. The financial means at their disposal are small. They concentrate on goods and commodities for everyday use. Small-scale production is based on forms of production which still represent "manufacture", i.e. crafts and technical crafts; in these countries, manual skills and corresponding capabilities predominate. Products

produced frequently have quality deficits or are manufactured with wide error toler-
ance margins, and thus are only suitable for sale in their country of origin.

For Central and Eastern European countries, pressure to adapt due to market forces
is a new challenge, further intensified by foreign competition. In some cases the
managers of firms have been able to make autonomous, market-oriented decisions
and develop individual strategies for survival. In view of the high degree of uncertainty regarding markets and the shortage of resources, such strategies tend to be
based on improvisation skills and lead to relatively simple, not very technology-
intensive production structures and to small production volumes (Portratz/Widmaier
1995).

To a large extent foreign investments in Central and Eastern European countries still
characteristically involve the manufacturing of large quantities of "low-quality"
products using technology- and capital-intensive production systems, or piecework
production/production on a subcontracting basis. In industrialized countries in the
West, small and medium-sized firms which are innovation-oriented or capable of
innovation play an important role in economic development; in most Central and
Eastern European countries, however, these firms are still lacking today, as an after-
effect of ownership law under former socialistic regimes and a previous lack of societal acceptance. Therefore in Central and East European countries the re-
structuring of the industrial sector is still incomplete. Whereas on the one hand the
survival and start-up chances of firms are still very hazardous, the structures of
suppliers at the various stages of the value-added chain also still appear underdeveloped, as do important related suppliers of pre-products and industrially-oriented
research.

In general terms, this means that at present the private sector is virtually unable to
fulfil tasks in the field of research and development (R&D). Moreover, industrial
R&D in the countries of Central and Eastern Europe was previously mainly concerned with carrying out adaptation developments, and was not oriented towards
innovative products or new production technologies. Industrial R&D potential was
previously one-sided in its qualification, and in the meantime it has suffered to
some extent from the fragmentation of industrial complexes and from cuts in personnel. For firms whose survival strategies do not lie in the area of sophisticated
technologies, future prospects are at best uncertain.

Technological R&D potentials do exist, however, in the public research institutions. In Central and Eastern European countries the performers of research were primarily the universities, public research institutes and academies (Schimank 1995). Due to the intensive research taking place in state institutions, the supply of qualified R&D results was comparatively large. Today, many Central and Eastern European countries still possess a broad range of research institutions. Some of these countries cultivate a presence in various different areas of basic research. For them, orientation towards the international scientific community both was, and is, a priority consideration. The universities and other scientific institutions regard themselves as an academic elite and, consequently, do not see themselves as "pre-thinkers" - or even problem-solvers - for industry, especially as at the present time private enterprise cannot be considered as a partner or client of the science sector.

Not only does the vertical structure of the research landscape separate industry from science; often its effects are also felt within the science sector itself (academy versus university). All in all, the exploitation of research potentials linked with industrial know-how in application-oriented research is thus too low.

Sociopolitical framework conditions

Policy and administrative law in Central and Eastern European countries do not provide incentives for innovation. Initiative, a willingness to bear risks and participate in free market competition are not sufficiently recompensed in terms of economic success. There is frequently a lack of generally valid regulations, particularly in the area of contract law, and a lack of (administrative) provisions for the legal enforcement of contracts in cases of conflict. The uncertainties with regard to these legal aspects constitute an obstacle to formal cooperation relations between actors, and negatively affect the subjective perception of success prospects for innovation projects in the private sector.

Often, policy regulations hardly allow for free communication or the free combination of resources. Policy is often centralistic, fairly inflexible and not very demand-oriented. Practically-oriented politico-administrative decisions are frequently impeded by a rigid adherence to the "letter of the law" in the implementation of regulations, and by long-drawn-out decision procedures. In the countries of Central and Eastern Europe, unlike countries in the West, it is not regarded as the self-evident duty of the state to make (scientific) knowledge available to the general public or industry. This situation is rendered all the more serious by the fact that in these countries, it is the state institutions which would be in the best position to initiate cooperative synergies between societal groups such as the science sector and industrial enterprises. Although in some Central and Eastern Europe countries the establishing of a new political order, including administrative and economic policy regulations is still not completed, in other countries many of the problems associated with transition have been overcome to a large extent.

It is in this complex, multi-faceted context that measures for an operative and strategic innovation and technology policy have to be developed for the individual countries in Central and Eastern European. This has to be accomplished in a situation of extreme shortage of resources and in the face of other urgent and pressing policy requirements (e.g. structural assistance for regions in need, payments to the unemployed, care of refugees). Thus for most of these countries, it would be generally true to state that since the beginning of the transition, an innovation or technology policy has existed only in a rudimentary form, if at all, and that existing innovation potentials are endangered.

Central and Eastern European countries should build up innovation-supportive relations between all relevant contributors of resources in society. These relations include the formal and informal exchange of information, supply and service networking and cooperations. The "mental gap" between science and industry has to be overcome in order to effectively exploit endogenous potentials. There must be greater awareness of the necessity to orient research more strongly towards the needs of industry. Up to now operational concepts have been lacking and cooperations have failed due to financial bottlenecks of the enterprises. The utilization of technological research results for the development of products which are mature for the market necessitates cooperation between science and industry, with relations taking the form of an intensive two-way exchange in which users of technological knowledge test out its suitability for industrial manufacturing, and the necessary modifications are made in a process of mutual learning.

3.2 The Implementation of Networks

The successful implementation of a network concept in Central and Eastern European countries will not necessarily result from the transfer of measures that have proved successful in other countries. The same activities may have very different impacts when applied under different specific societal and political framework conditions. However, it is possible to identify success factors that are independent of any specific system and adapt them to different societal conditions (Walter 1992). This should be borne in mind when transferring experiences to Central and Eastern European countries and in the implementation of transition assistance by Western countries. Networks have arisen in Germany and other Western industrialized nations over a relatively long time span, and mostly through trial and error. For reasons of time and economy of resources, a trial and error process is not suitable for countries in Central and Eastern Europe. Furthermore, due to the frame conditions described above, it cannot be assumed that in Central and Eastern Europe innovation networking between the various different sources of innovation potential will automatically occur.

Based on German experience, a possible procedure is now described for the infrastructurally-supported initiation and strengthening of networks at a national and regional level, with the possibility of integration into international innovation networks. This procedure takes account of already-existing institutional starting-points and relevant personal and political contacts in the country concerned, but also supports early self-organization of the networks.

State innovation and technology policy can support efficient, innovative cooperations in the form of "horizontal" networks between research and industry, if it succeeds in integrating the relevant actors into the networks: enterprises, research institutions, universities and suppliers of innovation services. This can be done by strengthening existing interactive relations and initiating new networks on the one hand, and on the other by identifying network deficits. If deficits are found, the missing network partners can be established by state initiatives as a part of innovation and technology policy.

The success or failure of innovation and technology policy measures supporting the network concept is decisively dependent on reaching a broad consensus of all relevant actors in policy, industry and science at an early stage (Koschatzky et al.

1995). It is also important that priority problem areas and fields of action should be jointly identified, and that concrete policy measures for support should be defined on this basis.

3.3   Examples of Networks in Central and Eastern European Countries

Networks support industrial innovation if they enable a demand-oriented exchange of techno-economic know-how to take place and mobilize funds for the production and market entry of new products and processes. First, suppliers of know-how can be networked with one another and with know-how users. The suppliers of know-how are primarily application-oriented research and development establishments, techno-scientific and economic institutions and training and higher education institutions, but also - insofar as they (still) exist - development departments and research groups in industry. The main users are enterprises of various sizes in different sectors. For activities to reflect real needs, there is a necessity for close cooperation between suppliers and users and for interactive supplier-user learning processes with alternation of roles. Networks in Central and Eastern European countries should involve all a country's regions and sectors of industry. For the mutual exchange of information and services to take place, spatial proximity of the actors is also important.

In the conditions that pertain in transitional countries, the responsibility for initiating and stimulating a network consisting of manufacturing firms, research establishments and services tends to lie with policy. The state should entrust specific tasks to network actors according to their specialist expertise, their capacities and their location or radius of action, and should partly finance these tasks in the initial phase. This is the point of application for innovation and technology policy instruments designed to support the expansion and formation of the network and promote cooperation activities between the network partners. As well as the institutional promotion of important institutions in the network, financial incentives will result in learning effects and will spur other initiatives - including private ones.

A strategy for the formation of an innovation network must bring in all relevant actors at various regional and national levels and in different industrial sectors. Although the networking relations that arise are between decentrally-active participants, it does appear important in the initiation phase to have a central instance which performs planning and coordination tasks on behalf of all the network partners.

An institution with an appropriate public contract can perform planning and coordination tasks in the network and can provide organizational support. However, this institution would not function as a centralistic planning body - rather, its importance would be in acting as a moderator in the generation of a regional or national modernization strategy and the formation of a consensus among all relevant actors in science, industry and policy. Network coordination requires techno-economic competence and an abundance of contacts to users and suppliers of innovation support services, in order to gather information about innovation services and the demand for them. Particularly important are data on industrial demand and the supply of technology, emerging technological and market trends. The institution, as the nodal point of the network, also takes on the documentation, acts as a contact partner for all other network partners and establishes active external contacts, for instance to international networks. An interface of this kind gives the network access to the direct use of globally available research results and, conversely, enables it to cooperate on equal terms in the international exchange of knowledge and know-how by making its own resources available. For Central and Eastern European countries, this would seem an important contribution towards integration into global networks and gaining a position in the international technology competition.

The decentral elements of the network structure to be established include public teaching and research institutions as well as industrial and sectoral associations. These can make their sectoral or specific knowledge and know-how available. Transfer and advisory offices can cover different specialist areas and contribute at a national level to a comprehensive, complementary offer. The abundance of highly specialized information contained in these institutions should be used by all network partners. As well as the technological input, surviving links and contacts to international science that may still exist in research are also important for the network.

Regional contact offices which are spatially accessible should be set up to provide users in the region with demand-oriented information and advice, and to mediate contacts. These offices should be run by existing industrial or technology-related institutions (e.g. chambers of commerce, technology centres) which, as actors at a regional level, have the advantage of intensive awareness and are well suited for organizing the exchange of specialist information. They are also in a position to make use of informal contacts in smoothing over, at a personal level, discrepancies that arise between network partners.

It is a good idea for the internal flow of communication and financial means within the network to be secured and organized by the coordinating institution. Also additional services should be provided such as the regular issuing of information letters, the organization of conferences and fairs to foster the exchange of experiences and the mediation of contacts. It seems to be important that communication is not "centralistic", but that all the actors intercommunicate. The contact offices also function as intermediaries, i.e. between the firms and suppliers of know-how. If the need arises, the network brings in other additional institutions. Existing gaps in the network are closed. Policy can support the establishing of appropriate institutions. Care should be taken to ensure their practical orientation, so that their services are accessible to all partners.

Since innovation and technology policy in Central and Eastern European countries is only able to implement measures involving relatively low financial resources, these measures should be directed towards the initiation of networks. Networks aim to mobilize and focus existing institutional and personal resources in order to strengthen industrial innovative activity, and stimulate firms which are as yet non-innovating to engage in innovative activities. Limited financial resources can be used for institutional funding to close gaps in networks, and for the promotion of specific cooperations between partners in the network. In this respect, public financial assistance should be regarded primarily as "initiation financing".

4. Help in the Initiation of Innovative Networks for Central and Eastern European Countries in the Process of Transition

The bringing together of existing resources in industry and science, and the initiation of innovative networks, is a difficult task under present conditions in Central and Eastern European countries. Thus it would appear important that developed industrialized countries such as Germany should offer these countries assistance in the process of transition, and should give "help towards self-help" to public organizations there. Assistance can support the planning of a policy of this kind by analyses, by the transfer of expert knowledge, by training and by advice during realization. A concept of this kind was developed by the Fraunhofer Institute for Systems and Innovation Research (ISI) at the request of the Federal Ministry of Education, Science, Research and Technology (cf. Figure 30). The concept is flexible and capable of adaptation to the transitional context of individual Central and Eastern European countries, as it integrates and supports existing structures and initiatives. The complex constellations of circumstances in these countries mean that any procedure used there has to be essentially exploratory. This implies that the time schedule, content and staffing requirements of the activities may have to be flexibly adapted to changing conditions.

Transitional assistance for Central and Eastern European countries aims to stimulate, in consensus with the partners, modernization processes based on the existing strengths of these countries, rather than giving (one-sided) policy recommendations whose mode of realization remains an open question. It is always essential in the "help towards self-help" approach that the decision-making and implementation activities should lie within the Central and Eastern European countries themselves. This requires strong autonomous motivation in these countries and commitment to a mode of realization by "small steps". However, a gradual, interactive procedure of this kind enables learning effects to take place and allows for a flexible demand orientation. The governments in Central and Eastern European countries themselves have the responsibility for the individual steps and for their coordination; they also provide the relevant services and funds.

Figure 30:   **Concept for help in the process of transition**

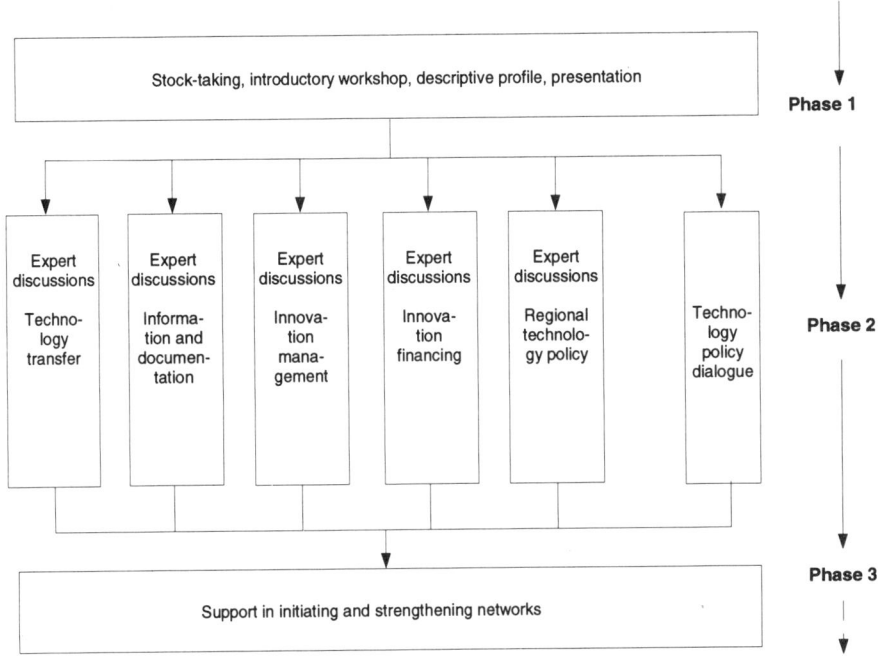

Transitional assistance by ISI (cf. Walter 1995) generally begin with an analysis of existing information in the form of a compact descriptive profile of the country concerned. The data on which this profile is based are collected and interpreted in the areas of policy, industry, science and regional structure. The compact profile serves as a basis for a more detailed inventory of the initial situation, for which information is gathered in interviews with experts in the areas named above. It is also a good idea to hold an introductory workshop in order to discuss at an early stage, with actors from politics, industry and science in the country concerned, the aims of the cooperation and the steps to be taken subsequently. A workshop can also serve as an opportunity for a brief outline and transfer of basic information on modern technology and innovation policy. It also creates an opportunity for the first exchanges of experience among specialists, which lead to personal contacts between all participants and are very helpful for future cooperation and the formation of a broad consensus. Subsequently, to record and evaluate the situation, interviews in the areas relevant to innovation are held with experts in ministries, universities and research institutes, enterprises, chambers of industry and commerce, banks and industrial associations. Presentation of the most important results of these interviews serves as an orientation aid in evaluating the initial situation with regard to providing help in transition. At a second workshop, the results of the inventory and first proposals for improvements can be presented to actors in policy, industry and science in the relevant country and intensively discussed with them. If the proposals meet with broad acceptance, they can be followed up by specialist meetings oriented towards implementation in those areas in which the inventory has revealed the greatest difficulties and information deficits.

Experts and practitioners in the areas of policy, industry and science concerned are invited to these implementation-oriented specialist meetings, together with speakers from abroad, with the purpose of defining problems and possible solutions more precisely. These meetings are intended for the intensive discussion and coordination of measures for improvements in specific areas, and the identification of possibilities for their realization. The papers and results of discussions are documented and passed on to interested parties. As a result of this comprehensive documentation, interested parties from policy and industry have direct access to international experience in their own strategic and project development.

In parallel to the specialist meetings, a technology policy dialogue should begin between policy-makers and experts from the West and the Central or Eastern European countries. Specialists from industry and science can discuss results so far and make appropriate suggestions for innovation and technology policy in view of the country's specific current situation. After the joint elaboration of suitable concepts follows the initiation and expansion of a network for the support of industrial innovation. Since in Central and Eastern European countries personnel with the necessary qualifications are only available to a limited extent and for particular tasks, great importance attaches to the focusing and organizational concentration of know-how, possible associated synergy effects and "short cuts". Compared with these aspects, possible conflicts of interests between the participating actors and their various spheres of competence appear secondary.

The last part of the transition assistance involving the initiation and strengthening of institutions has a different quality and requires more binding commitment than the preceding activities. For the networks, new institutions have to be created and the main topics covered by existing institutions have to be expanded or redefined.

Transitional assistance given by Western industrialized nations thus needs to be competent in specialist fields and able to hold its own against short-term, day-to-day problems. As well as advice and planning assistance, help may include training or study visits abroad for people from Central and Eastern European countries. This phase, involving the institutional realization of measures for improvement, is decisive for the success of transitional assistance as a whole: here the experience gained in the surveys and specialist discussions can be visibly concretised and can lead to improvements in the priority problem areas of industrial innovation.

All in all, transitional assistance for Central and Eastern European countries is characterized by numerous, parallel tasks with different time horizons, fluctuating determinant parameters and changing contact partners, frequent changes of situation and limited endogenous resources for action. This constellation overlays a conservative basic structure with a multitude of personal dependencies, resulting in low flexibility and mobility. Thus on the one hand there is a necessity for a long-term, integrated approach in transitional assistance, with gradual realization in successive steps and the possibility of correction; on the other hand, there is also a need for relationships of personal interchange, the use of changing procedures and powers of improvisation.

5. German Transitional Assistance for Slovenia and Croatia

As a consequence of the structure of the science system in former Yugoslavia and the subsequent independence of Slovenia and Croatia, a re-orientation of economy and science in Slovenia and Croatia was necessary. For industry to be able to exploit the existing research potential as soon as possible and at the lowest possible cost, science, industry and policy had to cooperate more intensively, both nationally and internationally. The German side was very interested in techno-scientific cooperation and in further developing the already existing relations between Germany and these countries. For this reason, Germany was providing transitional help for these two countries (Komac/Krawczynski 1994; Svarc/Lange 1995).

Figure 31:   **Map of Slovenia and Croatia (overview)**

Source: Information Service of the Frankfurter Allgemeinen Zeitung, August 1995

Due to the cooperation deficits between science and industry existing in Slovenia, a technology transfer network with complete coverage by regional contact offices appears an advantageous strategy. The central institutions of this network should concentrate on the initiation and execution of technology transfer projects and should assume responsibility for supporting the activities associated with the network (for example by networking with the relevant institutions). The aim is to support the development and market entry of new products and processes by the provision of financial assistance, by the transfer of know-how, by cooperation between research and industry and by techno-economic information and advice.

There are already a number of organizations such as technology transfer offices of universities, technology and incubator centres and advisory institutions in Slovenia. These can participate in the transfer network, taking on tasks related to their own field and retaining their independent status. At least to begin with they should receive some basic funding, but they should also aim at an early stage to earn their income from services to industry and science. In the example of a network structure described above, it is envisaged that a central nodal point should support the link between informative, advisory and research institutions and industry, and should also work in conjunction with regional contact offices to ensure that the needs of Slovenian industry are met by a comprehensive supply of innovation services covering all regions. All activities are aimed at the fast transfer of results from research to application, thus strengthening economic potentials.

In Croatia, the foreground is occupied at present by stabilizing the economy and combating the after-effects of the war. In consideration of the long-term perspective, however, the Croatian side has formed the intention of making a start now on the practice-oriented reorganization of research and development and the implementation of measures to initiate and strengthen technology transfer. Cooperation between science, industry and public administration has been underdeveloped so far. There are no institutions which could accelerate technology transfer. Thus transitional assistance by Germany should attempt to find ways of substantially improving the transfer situation. Proposals developed as a part of transitional assistance aim to strengthen cooperation between science and industry by the development, further expansion and foundation of institutions such as technology transfer offices, technology and incubator centres and technology parks. The Croatian side has made the decision to begin with the foundation and start-up of a technology transfer centre at Zagreb University, which can later also function as the nodal point for a nation-wide network.

The German Federal Government supported both these countries as a part of its transitional assistance to Central and Eastern European countries. In doing so, it follows a "help towards self-help" orientation, as described above. Work in the two countries demonstrates the need to adopt a flexible approach. In Slovenia, following the stock-taking phase and realization-oriented Slovenian-German specialist meetings concentrating on areas such as technology transfer, information and documentation, innovation management and innovation financing, the outcome was a bilateral Slovenian-German technology policy dialogue in which a broad consensus was reached on a further course of action. In Croatia, the stock-taking phase was followed by the decision to set up the technology transfer centre.

6. Looking Ahead

The Central and Eastern European countries bordering on Central Western Europe can only achieve international competitiveness if they have an innovative national economy. Other competition factors such as low labour costs and raw material resources do not offer a sustained strategy for integration into the international division of labour. In the long term, help for Central and Eastern European countries in the process of transition serves to assist the stabilization of democracy and the introduction of a social market economy. In the transition from one system to another, innovation research and innovation and technology policy in Central and Eastern European countries are confronted by contradictory expectations. Demand there is primarily oriented towards short-term support in building up market-driven innovation systems; however, this seems not to be possible without long-term planning. Assistance by western industrialized nations to countries in Central and Eastern Europe can only be productive of success if it is adapted to the specific frame conditions in these countries, which differ from conditions in the West. The specific circumstances of transition can only be taken sufficiently into account by performing an adequate empirical stock-taking, by defining demand on this basis and by adopting a procedure that is consensus-based. For both sides, transitional assistance is explorative, sensitive to feedback and change, and has long term horizons.

Experience from individual countries can be incorporated into transitional assistance. Thus transnational elements are integrated into Germany's transitional assistance to Slovenia and Croatia. These include the initiation of expert discussion and workshops for international participants and the organizing of joint training sessions in Germany.

The network approach illustrated here demonstrates concrete possibilities for making an effective contribution to the economic development of Central and Eastern European countries and helping to build up their international competitiveness, while taking account of their institutional situation and their specific strengths. Network theory offers concrete starting-points for the promotion of a cooperative development in these countries which targets the activation, focusing and complementation of existing potentials. In both countries, the first steps towards implementation have already been taken. There is a need for further research on the practical application of this approach in other Central and Eastern European countries. In particular, there is a need to develop indicators specifically suitable for the transitional situation, which can be used to identify and evaluate regional innovation potentials.

7.   Bibliography

Håkansson, H. (1989): Corporate Technological Behaviour. Co-Operation and Networks. London, New York.

Herden, R. (1992): Technologieorientierte Außenbeziehungen im betrieblichen Innovationsmanagement. Heidelberg.

Kline, S.J./Rosenberg, N. (1986): An overview of Innovation. In: Landau, R./Rosenberg, N. (Eds.): The Positive Sum Strategy. Washington, 275-305.

Komac, M./Krawczynski, J. (1994): Conceptual Approaches to the Support of industrial Research and Development in Slovenia (Workshop Proceedings). Forschungszentrum Jülich. Jülich.

Koschatzky, K./Gundrum, U./Muller, E. (1995): Regionale Innovations- und Technologieförderung. Ansatzpunkte für die Nutzung regionaler Innovationspotentiale. Working Paper. ISI: Karlsruhe.

Koschatzky, K./Breiner, S./Bördlein, R. et al. (1993): Technologieprofil der Region Rhein Main. Hauptstudie, Teil I zum Projekt High-Tech-Unternehmen in der Region Rhein Main im Auftrag des Umlandverbandes Frankfurt und der Wirtschaftsförderung Frankfurt GmbH. Frankfurt, Karlsruhe.

Landtag von Baden-Württemberg (1993): Wirtschaftsnahe Forschung in Baden-Württemberg - Antwort der Landesregierung auf die Große Anfrage der Fraktion der SPD, Drucksache 11/2449. September 1993. Stuttgart.

Meyer-Krahmer, F./Kuntze, U. (1992): Bestandsaufnahme der Forschungs- und Technologiepolitik. In: Grimmer K./Häusler, J./Kuhlmann, S./Simonis, G. (Eds.): Politische Techniksteuerung. Opladen, 95-117.

Potratz, W./Widmaier, B. (1995): Industrielle Perspektiven in Osteuropa: Industriepolitik oder Industrieentwicklung (Conference Paper). Tagung der Friedrich-Ebert-Stiftung: Osterweiterung der Europäischen Union am 7./8.7.1995 in Leipzig. Bonn

Schimank, U. (1995): Die Transformation der Forschungssysteme der mittel- und osteuropäischen Länder: Gemeinsamkeiten von Problemlagen und Problembearbeitung. In: Mayntz, R./Schimank, U./Weingart, P. (Eds.): Transformation mittel- und osteuropäischer Wissenschaftssysteme. Länderberichte. Opladen.

Svarc, J./Lange, S. (1995): Conceptual Approaches for an Industry-Related Promotion of Research an Development in Croatia. (Workshop Proceedings) Forschungszentrum Jülich. Jülich.

Sydow, J. (1992): Strategische Netzwerke. Evolution und Organisation. Wiesbaden.

Walter, G.H. (1992): Integration einheimischer Hochschulen in die industrielle Modernisierung der Dritten Welt. Bollschweil bei Freiburg.

Walter, G.H. (1995): Slovene-German Cooperation in the Field of Technology Policy. In: Pejovnik, S./Komac, M. (Eds.): FORUM Bled, International Scientific and Technological Cooperation: Problems, Challenges, Opportunities. Ministry of Science and Technology, Ljubljana.

List of Authors

Ulrike Broß

degree in business administration in 1994, studied at University of Cologne and Università Commerciale Luigi Bocconi, Milan. Research fellow in an EU-TACIS project and consultant in the public sector economics with Kienbaum Unternehmensberatung GmbH, Düsseldorf. Research assistant in the Finanzwissenschaftliches Forschungsinstitut, Cologne. Since November 1995 research fellow with Fraunhofer Institute for Systems and Innovation Research, Karlsruhe in the department "Innovation Services and Regional Development". Research areas: regional innovation systems (with focus on countries in transition) and innovation financing.

Uwe Gundrum

studied political and public administration science at the University of Constance (main topic: regional planning and infrastructure). Head of social planning and citizen participation department at the Office for Urban Renewal of the City of Cologne. Since 1980 public relations officer and scientist at the Fraunhofer Institute for Systems and Innovation Research (ISI), Karlsruhe (Germany). Research topics: Regional innovation and technology policy, regional conditions and impacts of new technologies (especially new information and communication technology), applications of new technologies in public administration, regional innovation processes and strategies.

Joachim Hemer

studied Electrical Engineering, Economics and Industrial Engineering at the Technical University of Darmstadt, Germany; Diploma 1977; 1977-1979 member of a research team at the Technical University of Darmstadt; 1980 project manager at the market research institute Marplan-Töpfer in Rodgau/Germany; since 1980 scientist with the Fraunhofer Institute for Systems and Innovation Research, Karlsruhe. Selected research areas: economic effects of sector interrelations, technological promotion and regional policy, market potentials and application of new technologies, technology-induced novel services and their economic interrelations, promotion and financing of technology-oriented small and medium-sized enterprises and their development processes, innovation financing.

Knut Koschatzky

studied geography and economics at Berlin's Free University and at Hanover University. 1986 PhD in economic geography. Served as assistant at the Department of Economic Geography and of the President of Hanover University. Joined Fraunhofer Institute for Systems and Innovation Research in 1988. During the same year scientific administration at the Bavarian Ministry of Economic Affairs. Head of the department "Innovation Services and Regional Development". Special research areas: regional innovation systems, regional technology policy, technology and innovation indicators, innovation and patent services, biotechnology, sensors.

Marianne Kulicke

studied business administration at the University of Saarbrücken; PhD in 1986; joined Fraunhofer Institute for Systems and Innovation Research in 1985. Deputy head of the department "Innovation Services and Regional Development". Special research areas: founding process and growth patterns of new technology-based firms, financing instruments and consulting offers for such firms, public programmes for the stimulation of such firms in Germany and other European countries, venture and seed capital markets.

Emmanuel Muller

studied Financial Management at the University of Strasbourg; 1993 Financial Management diploma, 1992-1993 research assistant at the Bureau d'Economie Théorique et Appliquée (BETA, Louis Pasteur University, Strasbourg), since 1993 Fraunhofer Institute for Systems and Innovation

Research, Karlsruhe. Special Research areas: regional and technology economics, venture capital.

Franz Pleschak

studied Engineering and Economics at the Technical University of Dresden; PhD in 1967 and post doctoral thesis in 1971. In 1970 he became a lecturer and in 1982 a full professor of Business Economics at the Technical University of Dresden. Joined Fraunhofer ISI in 1992, Joint Research Office on Innovation Economics at TU Bergakademie Freiberg (Saxony). Head of project for the scientific analyses within the pilot scheme "Promotion of New Technology-Based Firms in the New Federal States". Special research areas: R&D management, product and technology innovations, economic aspects of automation, CAD/CAM/CIM, firm foundations, technology centres.

Günter H. Walter

gained a Master of Business Administration (with technical background) at the University of Karlsruhe and a PhD in Political Science (Dr. phil.) at the Free University of Berlin. Since 1972 at Fraunhofer ISI as researcher and project leader. Lecturer in technological development and technology policy at University of Karlsruhe. Research areas: National and regional technology and innovation policy, technology transfer from R&D institutes to industry, modernisation of central European countries in transition, innovation in small and medium-sized enterprises.

Henning Werner

studied industrial engineering at the Technical University of Darmstadt. Entrance to Fraunhofer Institute for Systems and Innovation Research at the beginning of 1995. Special research areas: Technology-Based Foundations, Innovation Financing.

Udo Wupperfeld

studied economics at Mannheim University. 1996 PhD in business economics. From 1991 to 1996 at Fraunhofer ISI at the department "Innovation Services and Regional Development". Since 1996 Professor at the Fachhochschule Pforzheim and project leader at the Steinbeis-Transferzentrum Marketing, Logistik und Unternehmensplanung (MLU). Main areas of research and teaching, consulting: marketing, sales, new media (e.g. internet) and financing.

TECHNOLOGY, INNOVATION and POLICY

Series of the Fraunhofer Institute
for Systems and Innovation Research (ISI)

Volume 1:
Kerstin Cuhls, Terutaka Kuwahara
Outlook for Japanese and German
Future Technology
1994. ISBN 3-7908-0800-8

Volume 2:
Guido Reger, Stefan Kuhlmann
European Technology Policy
in Germany
1995. ISBN 3-7908-0826-1

Volume 3:
Guido Reger, Ulrich Schmoch (Eds.)
Organisation of Science and Technology
at the Watershed
1996. ISBN 3-7908-0910-1

Volume 4:
Oliver Pfirrmann, Udo Wupperfeld and
Joshua Lerner
Venture Capital and New Technology
Based Firms
1997. ISBN 3-7908-0968-3

Printing: Weihert-Druck GmbH, Darmstadt
Binding: Buchbinderei Schäffer, Grünstadt